P9-EDZ-972

AN INTRODUCTION TO PRACTICAL INFRA-RED SPECTROSCOPY

A. D. CROSS
B.Sc., Ph.D.
Research Laboratories, Syntex, S.A.,
Apartado 2679, Mexico, D.F.
Formerly Lecturer, Imperial
College of Science and
Technology, London

SECOND EDITION

LONDON
BUTTERWORTHS
1964

ENGLAND:	BUTTERWORTH & CO. (PUBLISHERS) LTD. LONDON: 88 Kingsway, W.C.2
AFRICA:	BUTTERWORTH & CO. (AFRICA) LTD. DURBAN: 33/35 Beach Grove
AUSTRALIA:	BUTTERWORTH & CO. (AUSTRALIA) LTD. SYDNEY: 6/8 O'Connell Street MELBOURNE: 473 Bourke Street BRISBANE: 240 Queen Street
CANADA:	BUTTERWORTH & CO. (CANADA) LTD. TORONTO: 1367 Danforth Avenue, 6
NEW ZEALAND:	BUTTERWORTH & CO. (AUSTRALIA) LTD. WELLINGTON: 49/51 Ballance Street AUCKLAND: 35 High Street
U.S.A.:	BUTTERWORTH INC. WASHINGTON, D.C.: 7235 Wisconsin Avenue, 14

First Edition 1960

Reprinted 1962

Second Edition 1964

Suggested U.D.C. Number: 543·422·4

Printed in Great Britain by Page Bros. (Norwich) Ltd.

PREFACE

FEW organic chemists engaged in molecular structural research or analytical control can now avoid calling upon infra-red absorption measurements for guidance at one time or another. Whether they have to make the measurements themselves or receive spectral charts of their compounds recorded by technicians, it is vital that they should have sufficient understanding of what is involved if they are to assess the results and make the correct interpretation. Here, too, it is not only a question of knowing what can be done by this method, but also what it cannot do and where real ambiguities can arise which must be solved by using other physico-chemical techniques.

A number of authoritative books and articles now exist about all this work, but this little manual prepared by Dr. Cross will none the less be useful, since it outlines concisely all the essentials of the theory, measurements and applications required by the majority of organic chemists. He has collected together a well chosen set of standard reference data and set it out in a readily usable form. I am sure that the book will interest a wide circle of users and undergraduate readers.

H. W. THOMPSON

St. John's College, Oxford
November 2nd, 1959.

ACKNOWLEDGEMENTS

SEVERAL major works already exist which provide comprehensive surveys of the literature, theory and practice of infra-red spectroscopy. They contain a wealth of detail and references to which all others interested in this subject must inevitably turn. I wish to acknowledge the great help derived from Dr. Bellamy's masterful book, the text of Jones and Sandorfy, and the earlier works of Randall, Fowler, Fuson and Dangl, and of Barnes, Gore, Liddell and Van Zandt Williams. I am deeply indebted to Dr. E. S. Waight and Dr. C. J. Timmons for invaluable advice during the preparation of the manuscript and to Dr. L. J. Bellamy, Dr. G. Eglinton, Dr. G. W. Kirby and Dr. J. K. Sutherland who subsequently read and constructively criticized the script. I record here my thanks to Professor S. Shibata for information on Japanese spectrophotometers; to the many research spectroscopists who spared time to discuss problems with me and to demonstrate their various infra-red instruments; and to all the instrument manufacturers for their generous provision of technical and scientific information.

My sincere gratitude is owing to Dr. H. W. Thompson for his Preface. My final thanks are reserved for my mentor, Professor D. H. R. Barton who, besides contributing a Foreword, offered continuous personal encouragement.

A.D.C.
July 1959

CONTENTS

FOREWORD

THE past two decades have witnessed the arrival of the infra-red spectrophotometer as an instrument of outstanding usefulness in all branches of chemistry. Already in widespread application in research, analytical and industrial laboratories, it is being belatedly introduced into the Undergraduate's course of study—a development which will undoubtedly spread with the arrival of robust low-cost instruments. It is primarily to aid the organic chemist, unfamiliar with practical infra-red spectroscopy, that this book has been written, though it should serve too as a convenient summary of techniques and data for those more familiar with the field.

Over the past few years Dr. Cross has introduced many newcomers to the general operating procedures and sampling techniques of infra-red absorption spectroscopy. He has shown that, given comprehensive initial instruction, and adequate supervision subsequently, students may operate instruments very satisfactorily and with great benefit to their understanding of the subject and to the progress of their work. Such experience prompted his opinion that these practices should be adopted more generally and that an Introduction to the field outside of review articles and special publications would be helpful. The present book meets these requirements in a very satisfactory manner. I am glad to recommend it to all students of Organic Chemistry and to others who may wish for an initiation into the practical applications of infra-red spectroscopy. In addition, the numerous tables provide welcome reference data in easily available form. The modest price of this book is another feature which is highly commendable.

Imperial College, London D. H. R. BARTON
October 26th, 1959.

AUTHOR'S NOTES ON SECOND EDITION

FOUR years have passed since the original text was brought to completion. The decision to prepare a second edition, rather than continue reprinting the first, was motivated by several factors, of which the marked progress in instrumentation should be mentioned. Relatively inexpensive grating spectrophotometers are now available, and the table on currently-available machines shows substantial additions to that compiled earlier.

The days when infra-red intensity measurements will be made routinely in the laboratory, although nearer, are still distant. Indeed, the advent of nuclear magnetic resonance spectroscopy with its manifold applications has undoubtedly offered in certain cases an easier, more accurate, alternative method for obtaining equivalent information to that derivable by quantitative infra-red spectroscopy. This advance is to be welcomed, not resisted. The chemist should freely depend upon all available forms of spectroscopy, not on one alone.

Most of the major revisions of the text fall in Part I and reflect the many developments and refinements in practical technique. I have been able to correct or clarify the text where appropriate.

I am indebted to the constructive critics of the First Edition, as well as to many friends, for suggestions as to how improvements were possible.

A.D.C.
November 1963

PART I

INTRODUCTION

THE first half of the book is devoted to the simple theory and practical aspects of infra-red spectroscopy. The emphasis, throughout, is on the qualitative rather than quantitative analytical value of infra-red, since it is with the identification of specific functional groups that the organic chemist is mostly concerned.

A relatively small section on theory is included since this is adequately dealt with elsewhere[1–4]. This is followed by a brief summary of the various uses of infra-red spectroscopy. The next section deals with the spectrophotometer itself and includes information on machines currently available or to be available shortly. Discussion is limited to double-beam instruments since their single-beam counterparts are less convenient for routine analysis in organic chemistry research laboratories.

Cell construction and sample preparation are considered in detail, as are also the importance of correct selections of prisms or gratings and of phase and sample diluents; these choices present valuable flexibility in instrument operation. The advantages and limitations of these variables are elaborated and the calibration procedure for the ultimate spectrum is included.

A separate section is devoted to problems of intensity measurements and the information to be gained from them. Hydrogen bonding is of especial interest in infra-red studies and the penultimate section deals with this phenomenon. Finally an outline is given of the procedure for interpretation of a spectrum.

Throughout, attention is drawn repeatedly to sources of error and to limitations of the instrument, materials and operator. Following the recommendations of the Royal Society Committee, wave numbers, rather than wavelengths, are referred to in the text. The argument for general adoption of frequency with the wave number scale has been briefly outlined by JONES and SANDORFY[5]. However, since many results are still published using wavelengths, the corresponding values on this scale are given in parentheses throughout the text and in the tables of both Part I and Part II; a table of reciprocals is also provided. Several methods for presentation of intensity measurements are currently in use, but there is an urgent need for an internationally agreed system of units for the presentation of results.

Important references are given throughout the text but comprehensive collections of relevant references, whose inclusion would substantially increase the size and cost of this book, are available elsewhere[1–9].

ELEMENTARY THEORY OF INFRA-RED SPECTROSCOPY

INFRA-RED radiation promotes transitions in a molecule between rotational and vibrational energy levels of the ground (lowest) electronic energy state*.

In a simple diatomic molecule A—B the only vibration which can occur is a periodic stretching along the A—B bond. Stretching vibrations resemble the oscillations of two bodies connected by a spring and the same mathematical treatment, namely Hooke's law, is applicable to a first approximation. For stretching of the bond A—B, the vibrational frequency ν (cm^{-1}) is given by equation (1)

$$\nu = \frac{1}{2\pi c}\left(\frac{f}{\mu}\right)^{\frac{1}{2}} \quad . \quad . \quad . \quad (1)$$

where c is the velocity of light, f the force constant of the bond, and μ the reduced mass of the system, as defined by equation (2)

$$\mu = \frac{m_A . m_B}{m_A + m_B} \quad . \quad . \quad . \quad (2)$$

where m_A and m_B are the individual masses of A and B.

Stretching vibrations of individual bonds within more complex molecules may be considered similarly, though other vibrations become possible and absorption band frequencies are influenced by other factors (vide infra). Substitution[10] in equation (1) of accepted numerical values of c, f and μ for the C—H bond gives a frequency of 3,040 cm^{-1} (3·29 μ), which is in tolerable agreement with found bond frequencies of 2,975–2,950 cm^{-1} (3·36–3·39 μ) and 2,885–2,860 cm^{-1} (3·47–3·50 μ) for methyl group C—H stretching vibrations. Such calculations are of greatest value when the atoms joined by the bond have large mass difference and when the remainder of the molecule exerts little influence on the bond motion under consideration. This condition is met when one atom is hydrogen and, in consequence, the A—H stretching mode frequencies are among the most thoroughly studied and valuable bands available for diagnostic purposes.

By convention, band positions are quoted in units of wave number (ν) which are expressed in reciprocal centimetres (cm^{-1}), usually styled band frequencies. However, the true unit of frequency ($\bar{\nu}$) is given in reciprocal seconds (sec^{-1}). An alternative term, wavelength (λ), measured in microns (μ) is also used to indicate band position. The relation between these units is given in the expressions

$$\nu = \frac{1}{\lambda}; \quad \bar{\nu} = \frac{c}{\lambda} \quad (c = \text{velocity of light})$$

$1\,\mu = 10^{-4}$ cm $= 10^4$ Å; hence, $10\,\mu = 1{,}000$ cm^{-1} and $2{\cdot}50\,\mu = 4{,}000$ cm^{-1}. Measurement of absorption band intensity is considered on p. 36. A complete review has been made of the presentation of absorption spectra

* This is in contrast to the energetically more powerful ultra-violet radiation, which facilitates transitions between vibrational and rotational energy levels of different electronic levels.

data, clarifying the superfluity of terms and units and recommending standardization[11].

A non-linear molecule of n atoms has $3n$ degrees of freedom which are distributed as 3 rotational, 3 translational, and $3n-6$ vibrational motions, each with a characteristic fundamental band frequency. However, since absorption occurs only where a change of the dipolar character of a molecule takes place, total symmetry about a bond will eliminate certain absorption bands*. Spectroscopists, in extending their studies from simple to non-

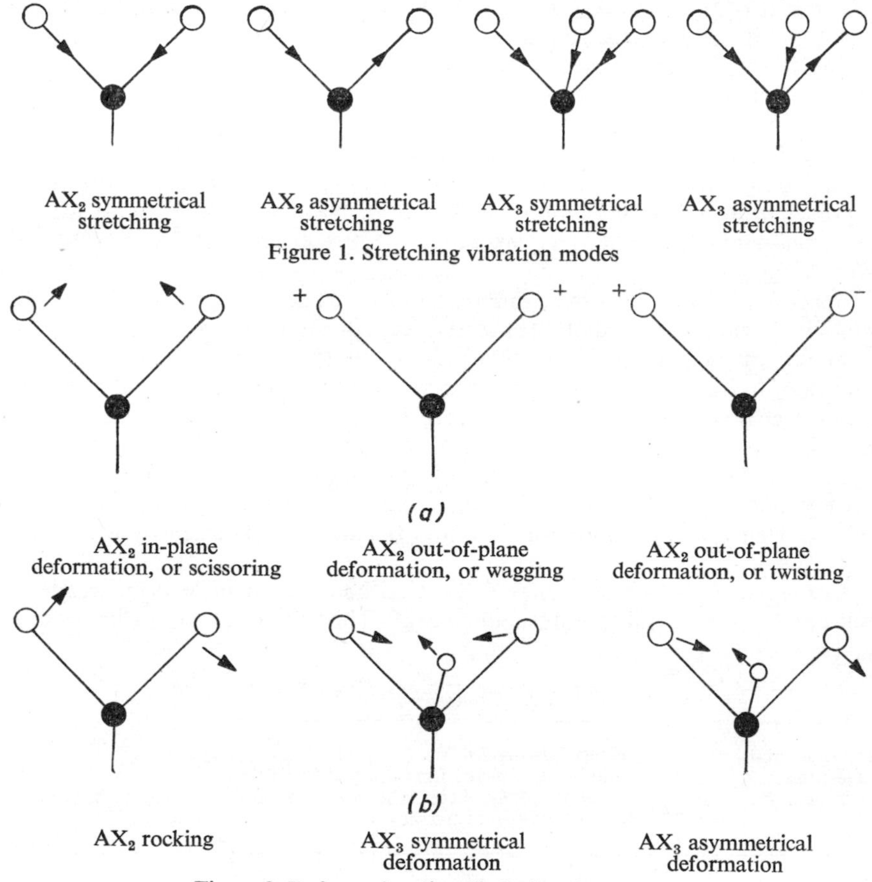

Figure 1. Stretching vibration modes

Figure 2. Deformation (bending) vibration modes†

* The band for C=C stretching is absent in the spectrum of *trans*-dichloroethylene, but for the same theoretical reason, the *cis* form should, and does, give a band specific for C=C stretching. Similarly, the symmetrical nitrogen and ethane molecules give no N—N or C—C stretching absorptions.

† Arrows indicate periodic oscillation in the directions shown. For a trigonal atom A, as in $B{=}AX_2$ (e.g. terminal methylene), these motions are in the plane of the three atoms involved (plane of the paper). When A is tetrahedral, as $—AX_2—$ in a chain (e.g. methylene in a paraffin), the arrows represent motions at right angles to the axis or general plane of the molecule.

The + and − signs represent, respectively, periodic motions above and below the plane of the paper, i.e. out-of-plane for the planar grouping $B{=}AX_2$, or along the axis of a chain molecule for $—AX_2—$.

linear, complex molecules, have verified that specific absorption bands for particular bonds or groups within a molecule occur at, or near, the expected frequencies. Fairly constant shifts of band frequency have been correlated with certain changes in structural or external environment. These results constitute the basis of modern qualitative analytical work in organic chemistry and are summarized in the correlation charts and tables which form Part II.

Bond vibration modes are divisible into two distinct types, stretching and bending (deformation) vibrations, the former constituting the periodic stretchings of the bond A—B along the bond axis. Bending vibrations of the bond A—B are displacements occurring at right angles to the bond axis. Many valuable stretching modes, particularly those involving hydrogen (A—H bonds), or of multiple bonds, occur at higher frequencies, 4,000–1,400 cm^{-1} (2·50–7·15 μ).

Several different types of vibrations have been defined; these are represented diagrammatically in *Figure 1* (stretching vibrations) and *Figure 2* (deformation or bending vibrations) for both AX_2 and AX_3 groups. Many absorption bands in the complex range 1,600–600 cm^{-1} (6·25–16·67μ) are still of unconfirmed origin.

Spectroscopists sometimes differentiate between the various vibration modes by using the symbol ν for stretching vibration frequencies, δ for frequencies of deformations involving bond-angle changes, τ for twisting vibration frequencies, and δ' (or π) for out-of-plane deformation or wagging mode frequencies. A further term, breathing, describes a completely symmetrical ring expansion and contraction vibration (C—C stretchings) in cyclic compounds. Assignment of band frequencies to specific vibration modes is often difficult and the cause of disagreement—three different interpretations of the origins of the Amide I and Amide II bands are assessed by RANDALL *et al.*[12].

Combinations of fundamental absorption band frequencies occur readily, but combination bands are normally weak*. The different types of bands are given in *Table* 1

Table 1. Absorption Band Types

Fundamental	Primary absorption band for each mode of vibration
Overtone	Band at multiple of fundamental band frequency
Combination bands	Bands near frequencies which are the sum or difference of two or more fundamental frequencies

For two fundamental frequencies, x and y, first overtones will occur near $2x$ and $2y$, second overtones near $3x$ and $3y$, and so on, and combination bands can appear at $x + y$ and $x - y$ cm^{-1}. All vibration mode absorption frequencies are susceptible, in varying degrees, to small alterations in the remainder of the molecule, and the highly characteristic nature of an infrared spectrum for every organic compound is therefore comprehensible.

Although bond elasticity, represented in equation (1) by the force constant

* Combination bands of pure compounds are sometimes misinterpreted as weak fundamental bands due to impurities. Thus the first overtone of the C=O stretching-band frequency may be erroneously ascribed to hydroxyl absorption.

f, and the relative masses of the bonded atoms constitute the two most important factors determining frequency, there are a host of other effects, both internal and external with respect to the molecule, which influence the absorption frequency. Electrical effects, steric effects, nature, size and electronegativity of neighbouring atoms, phase changes and hydrogen bonding may all cause shifts in frequency. Small changes of environment may sometimes be correlated with observed constant frequency shifts. Multiple bond and A—H bond stretching absorption frequencies are least affected by internal structural changes, except when intramolecular hydrogen bonding is involved, but are more susceptible to alterations in the external environment. Conversely, single-bond skeletal stretching vibrations between identical atoms, or atoms of similar mass, and the majority of bending vibrations are markedly influenced by internal structural changes. The less susceptible a group absorption frequency is to internal and external environmental changes, the more valuable it becomes for correlation purposes.

USES OF INFRA-RED SPECTROSCOPY

ALTHOUGH the organic chemist is most frequently concerned with the uses of infra-red spectroscopic data for identification of compounds, many other applications have been developed. Some of these require special adaptations of an instrument but the overall approach remains one of associating certain absorptions with specific groups within a molecule. Some of the different uses are outlined briefly.

Identification of a Substance

The infra-red spectrum of a compound is characteristic of that compound and may be used for identification, just as melting point, refractive index, boiling point, optical rotation, X-ray powder photograph and other physical constants are used. In comparative studies of two substances, therefore, identical infra-red spectra infer identical substances, with a few reservations. Spectral comparisons are normally made in dilute solutions, since a pure compound may crystallize in different forms, each with a characteristic solid-phase spectrum but giving identical dilute solution spectra. Furthermore, optical isomers show identical spectra in solution but racemates and enantiomers may give different spectra in the solid state owing to different packing arrangements within the crystal, and therefore no conclusions concerning the identity of enantiomers can be drawn. Solution spectra comparisons fail to establish identity for compounds containing large numbers of identical structural units, e.g. polymers or long aliphatic chains. In these cases addition or removal of a few structural units causes no observable alterations in the solution spectrum. However, solid-phase spectra are useful since different chain lengths require different unit cell dimensions within the crystal, with consequent changes in the solid-phase spectrum. Studies of polymer chains, fibres and crystals have been furthered by use of polarized infra-red radiation (p. 46). Hence, identity of both the solution spectra and the solid-phase spectra will establish identity. This method of proving identity is often superior to a mixed melting point, especially since a large number of characteristic bands is available for the comparison. Particularly careful scrutiny of the spectra becomes essential when the possibility of stereoisomerism exists. Compounds differing by some point of stereochemistry have different spectra, although the differences may not always be extensive. The following procedure for establishing identity has proved suitable.

(1) Use equal concentrations of pure substances and compare spectra in two different media, preferably in two different states, e.g. mull and solution or solid disc and solution.

(2) Obtain spectra at concentrations high enough to allow comparison of minor peaks. Comparisons are easily made by superimposing one spectrum upon the other and illuminating from beneath by a strong diffuse light, or by recording both spectra on the same chart.

(3) Compare intensities. If one spectrum is weaker than the other then all

peaks should display the same diminution of intensity. Such comparisons are best made on logarithmic-scale charts where linear comparison of absorptions becomes possible.

It is to be noted, however, that procedures 1–3 will not always be necessary since the first comparison of spectra, coupled with a mixed melting point, is often sufficient to establish identity beyond doubt. A direct consequence of identity is a 'straight line' spectrum when equal concentrations of both substances in carefully matched cells are placed in the reference and sample beams of the spectrophotometer.

Determination of Molecular Structure

A qualitative analytical approach to molecular structure, on the basis of infra-red spectra, is now possible. A separate section on interpretation of spectra, using the tabulated information of Part II, is included on p. 42. It may be stated here that in this type of work there is no substitute for experience, and it is of real advantage to the newcomer to examine many different types of compound. An exemplary use of spectral analysis has been the establishment of the lactam structure of 2- and 4-hydroxypyridines[13–15] and related compounds. Conversely the 2- and 4-aminopyridines and other heterocyclic amines[16] only show bands for the —NH_2 amino group, and no evidence of an imino form is apparent in the spectrum.

Determination of Purity, Quantitative Analysis and Production Control

These three procedures are considered together, since all involve, essentially, an estimation of the concentrations of several components of a mixture. In a direct determination of purity the technique is the same as proof of identity, if the pure compound is available for a reference spectrum. The presence of impurities will cause reduced sharpness of individual bands, a general blurring of the spectrum and the appearance of extra bands. High concentrations may be necessary to see the extra 'impurity' bands clearly. An approximate curve for the impurities is produced by a subtraction curve of the pure compound (reference beam) from the impure sample (sample beam of the instrument). This differential analysis procedure may allow identification of the impurities. The disappearance of bands specific for the impurity may be followed by spectral examination after successive purifications and their complete elimination is a useful criterion of purity. The above approach is also the basis of production control on an industrial or small laboratory scale. Here, the 'impurities' will consist of unchanged starting material and unwanted by-products of the reaction and a rough estimate of their concentrations may be made by intensity comparisons. In industry the yields required in a process may be an optimum rather than the maximum but in either case the development of an absorption band characteristic of the required product can be followed. A plot of its approximate intensity against time will reveal when no further increase in product concentration results (maximum yield); alternatively the reaction may be halted at a pre-selected product concentration for optimum yield.

A refinement of the impurity analysis procedure is the essence of quantitative analysis. Here, the analyst is required to estimate, as accurately as possible, the percentage concentrations of the components in a mixture. Using infra-red spectroscopy this determination rests solely on comparative absorption intensities, and the absorption intensities of the pure components must be known for at least one characteristic strong band in the spectrum of each component. Accurate measurement of the mixture concentration and of the thickness of the absorbing specimen then allows calculation of component concentration on the basis of Beer's law. Alternatively, each component is examined in its pure form at a series of concentrations and a calibration curve obtained of concentration against absorption intensity for a selected band. Using the same path length cell for the mixture, the concentration of the components may be obtained by measurement of the intensities of a characteristic band for each component, and reading from the relevant calibration curve the corresponding component concentration. This method obviates the necessity of measuring cell width (path length) by interferometric methods, and cancels errors due to non-parallel cell window surfaces, besides permitting quantitative analysis of substances which do not obey Beer's law (see p. 36).

A further factor which must be considered is the possible absorption by impurities at the same absorption frequency as the component. This complication may be resolved generally by study of two band intensities for the component.

Reaction Kinetic Studies

Rates of reaction may be measured in a number of ways. Consumption of starting materials and appearance of product are obvious points at which spectroscopy can be employed. By a simple mechanical device an infra-red spectrophotometer may be rigged to record a series of curves by repeated scanning over a pre-selected range at regular time intervals. A direct plot of absorption (per cent) against time is therefore available for any chosen band, whether it is an increasing absorption due to product formation or a disappearing absorption band of a reactant. Reaction kinetic studies are specific cases of quantitative analysis. Individual rates of consumption of several reactants may be determined and the existence of intermediates established. Rearrangements of allylic halides (I $\rightleftharpoons$ II) have been studied spectroscopically[17]. The intensity of the band for C—H out-of-plane deformation decreased with time when a pure sample of compound I was kept at 131° C; the equilibrium mixture ratio was calculated when no further intensity reduction occurred with time. Progressive diminution of the intensity of a band at 947 cm^{-1} for the pure primary halide (II) gave, under the same experimental conditions, an identical equilibrium mixture spectrum, consistent with the proportions 21 per cent I: 79 per cent II. Intermediate formation of the olefinic acid during the reduction of propiolic acid to propionic acid was similarly demonstrated by infra-red studies, since a band for =C—H stretching in acrylic acid developed and disappeared in the course of hydrogenation[18].

Very fast reactions are more difficult to study since instrumental limitations interfere; among these, the response time required from initiation of the signal by the detector to recording by the pen, and the reduced resolving powers of the instrument through scanning at high speeds may be serious factors. Several devices have been invented to surmount these practical difficulties[19].

Fundamental Studies of Molecules

Spectroscopists obtain fundamental data on molecular geometry by mathematical analysis of very accurate infra-red spectra, particularly of gases[1]. Only small molecules (2–8 atoms) in the vapour phase may be studied by this method. In a gas the molecules are free to rotate and a fine structure can appear in the absorption spectrum, corresponding to changes in rotational energy of the molecule which accompany vibrational transitions. This fine structure of an absorption band is observed only for small molecules where the individual rotational energy levels are sufficiently far apart to permit the main absorption band to be resolved. Larger molecules have individual rotational levels which are too close together for resolution, and only curves outlining the whole rotation–vibration absorption appear.

CONSIDERATION is now given to the many aspects of spectrophotometer construction and how the limitations of instruments in operation influence the quality of spectra. Infra-red spectrophotometers operate according to simple principles and their mechanical and electrical complexities are technical devices to transform minute energy absorption variations into an accurate spectrum recording.

Instrument Design

Several components are fundamental to every modern double-beam infra-red spectrophotometer. A source provides radiation over the whole infra-red spectrum; the monochromator disperses this light and then selects a narrow-frequency range, the energy of which is measured by a detector; the latter transforms energy received into an electrical signal which is then amplified and registered by a recorder. The whole light path and the ultimate focusing of the source image on the detector is determined by precision-manufactured mirrors. *Figure* 3 illustrates the optical path and principal

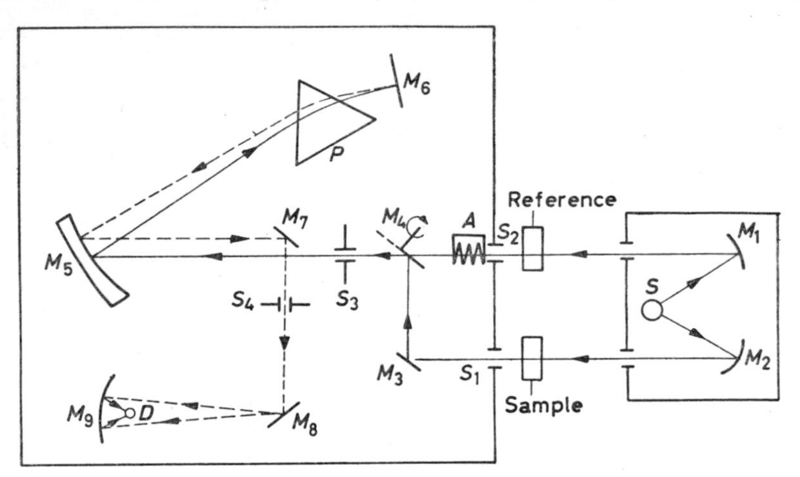

Figure 3. Simplified spectrophotometer

components of a hypothetical spectrophotometer. Light from the radiation source, S, is reflected by mirrors, M_1 and M_2, to give identical sample and reference beams. Each of these focuses individually upon vertical entrance slits, S_1 and S_2, the sample and reference cells being positioned in the two narrow beams near their foci. Transmitted light is then directed, by a mirror M_3, on to a rotating sector mirror (or oscillating plane mirror) M_4. The latter serves first to reflect the sample beam towards the monochromator entrance slits S_3 and then, as it rotates (or oscillates), to block the sample beam and allow the reference beam to pass on to the entrance slit. In this manner an image of the source, from alternating sample and reference beams, is focused upon the entrance slit, S_3. A collimating mirror, M_5, reflects parallel light to a prism, P, through which it passes only to be reflected back again through the prism by a rotatable plane (Littrow) mirror, M_6, and the prism disperses the light beam into its spectrum. Only a narrow range of the dispersed light collected by the collimator mirror becomes focused upon a plane mirror, M_7, which deflects it out through the monochromator exit slit, S_4. A further plane mirror, M_8, turns the light towards the condenser, M_9, which focuses it sharply upon the detector, D. When the energy of the light transmitted by both sample and reference cells is equal, no signal is produced by the detector. However, absorption of radiation by the sample results in inequality of the two transmitted beams falling on the detector and a pulsating electrical signal is produced, of the same frequency as the frequency of beam splitting by the sector mirror. This is amplified electronically and rectified, and used to move an attenuator, A, across the reference beam, cutting down the amount of transmitted light until an energy balance between sample and reference beams is restored (optical null method); at this point the detector ceases to emit a signal. The amount of reference-beam reduction necessary to balance the transmitted beam energies is a direct measure of the absorption by the sample. Synchronization of the attenuator with the recording pen gives a value of sample absorption as a pen trace on the paper chart. Strong absorption by the sample causes a large beam energy difference and, hence, a proportionately strong detector signal, to drive the attenuator well into the sample beam, nullify the energy difference and cancel the signal. When this position is reached both the attenuator and pen remain stationary. Rotation of the Littrow mirror changes the frequency of light reaching the detector, which may be accompanied by a change in sample absorption. If this is so, the whole sequence of signal and counteracting response is initiated and in this manner the whole spectrum is scanned continuously. The principle of beam balance ensures greater accuracy by elimination of errors due to variations in the radiation source, and the detector, besides providing the basis for differential analysis (a common component, e.g. solvent or impurity, in equal concentration in both sample and reference cells is effectively balanced out to zero absorption). A further advantage of double beam operation is the elimination of signal variations due to change of amplifier gain.

Design and function of the major instrument components have a significant influence on its versatility and operational accuracy. These variables are discussed below in greater detail.

Sources—Infra-red radiation is produced by electrically heating a Nernst filament (a high-resistance, brittle element composed chiefly of the powdered, sintered oxides of zirconium, thorium and cerium held together by a binding material), or a Globar (rod of silicon carbide) or any other suitable material. At a temperature in the range 1,100–1,800° C, depending on the filament material, the incandescent filament emits radiation of the desired intensity. Source diameters must be sufficiently large to provide an image wider than the entrance and exit slits at their maximum breadth. At lower frequencies (<600 cm^{-1}), using potassium bromide optics, the slits must be opened wide to allow more energy into the monochromator, since at these frequencies the energy emitted by the source is falling rapidly (*Figure 4*). Over the range 4,000–600 cm^{-1} the emitted energy intensity falls appreciably. Maximum energy radiation for a Globar is at 5,500–5,000 cm^{-1} (1·8–2 μ) while that of a Nernst is at 7,100 cm^{-1} (1·4 μ). Globars are of larger diameter and their emitted energy falls less (*ca.* 600 fold) on passing from 5,000 cm^{-1} to 600 cm^{-1} (2–16·7 μ) than occurs with a Nernst, where the fall in energy exceeds 1,000 fold; the Globar is more useful at lower frequencies.

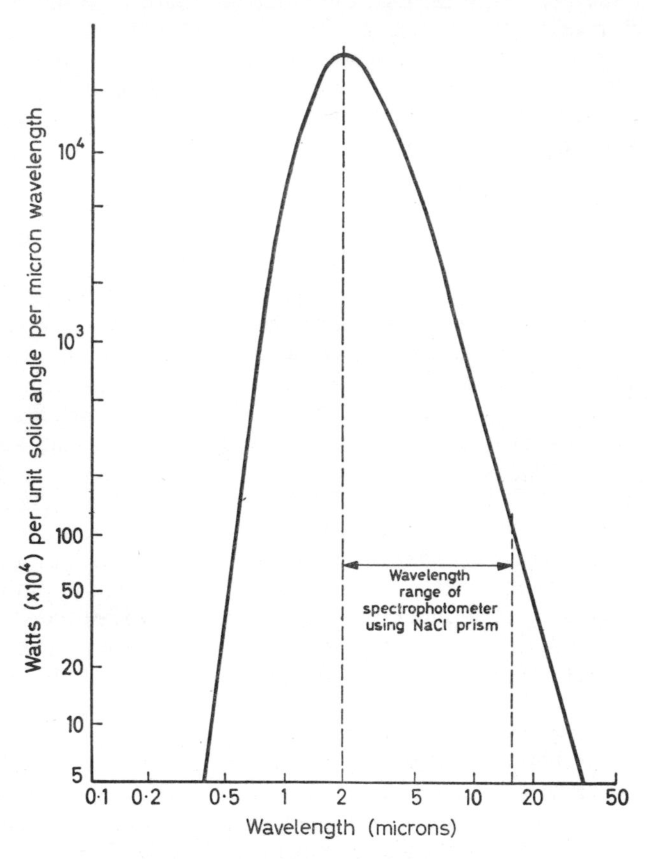

Figure 4. Distribution of radiation from a 1 cm square black body source at 1,100° K (Globar). (*By courtesy of* Baird-Atomic, Inc.)

Beam balance—It is important that the beams should be of equal energy prior to insertion of the sample and reference cells and auxiliary balancing attenuators are incorporated for this purpose. This balance should be checked regularly for all instruments.

Monochromators—The narrow source image which is focused on the monochromator entrance slit consists of undispersed infra-red radiation. A collimator reflects this beam as parallel light to a prism or grating which disperses the spectrum. Only a small portion of the dispersed light spectrum becomes focused by the collimator upon the monochromator exit slit. Narrow exit slits ensure that light of only a narrow-frequency range (approximating to monochromatic light) passes out of the monochromator. The dispersion of light by any prism material is directly dependent upon its refractive index which changes with light frequency though, even so, the dispersion properties of a prism are not constant with changing wave numbers. At least two different prisms are necessary to thoroughly cover the frequency range 4,000–650 cm^{-1}. A single rock-salt prism is often used as a compromise for the whole of this range but its dispersion decreases steadily towards 4,000 cm^{-1} and its deficiencies above 2,000 cm^{-1} (below 5 μ) should be noted. Sundry methods are available to increase the dispersion of the monochromator unit, and include very large prisms, multiple-pass single-prism units, multiple-pass multi-prism systems, gratings with a foreprism for spectral order selection and prism-grating double monochromators (see p. 35). The narrowness of the frequency range focused by the collimator on the exit slit is dependent upon both the dispersion by the prism or grating and the width of the entrance slit. The portion of the infra-red spectrum passing through the exit slit is determined by the Littrow mirror angle (or grating angle) and, by rotation of the Littrow mirror, the light reaching the exit slit changes frequency steadily until the whole spectrum has been scanned. A cam, carefully constructed, turns the Littrow mirror at a non-uniform rate of angular rotation to give a linear change in frequency, and its driving motor also controls automatic entrance and exit slit openings (vide infra) as well as the wave number counter. With a narrow entrance slit, only a thin beam of light is admitted and this is more efficiently dispersed than are broader beams. Similarly, narrow exit slits select a smaller frequency range from the dispersed spectrum for admission to the detector than do wide exit slits; exit or entrance slits are curved in outline to compensate for image curvature by the prism. Briefly, narrow slits, high-quality mirrors and units with good dispersion properties lead to an instrument with high resolution. One further major factor, the speed of scan, is considered on p. 15. Resolution (resolving power) is the ability of the spectrophotometer to separate light of any one frequency from light of closely similar frequencies.

Detectors—Light from the monochromator exit slit is focused upon a device which detects and measures the radiant energy by means of its heating effect. Detection of the exceedingly small temperature variations caused by radiant energy variations is most frequently accomplished by a bolometer or a thermocouple; a thermopile is a series of thermocouples linked together. In the bolometer the temperature rise due to radiation causes a change in electrical resistance which is used to vary a voltage. The thermocouple uses the radiant energy to heat one of two metal junctions and set up an electromotive force between the junctions, the voltage of which is directly proportional to the amount of radiant energy. The relatively slow response of both these detectors lengthens the overall time for an absorption difference to be signalled to the recording pen. The very weak electrical currents emanating from the detectors require manifold amplification to furnish a current capable of driving servo motors. Unicam Instruments Ltd. use a Golay pneumatic detector in their model SP 100. This consists of a gas-filled chamber which undergoes a pressure rise when heated by radiant energy. Small pressure variations cause deflections of one wall of the chamber; this movable wall also functions as a mirror and reflects a light beam directed upon it to a photocell, the amount of light reflected bearing a direct relation to the gas chamber expansion and, hence, to the radiant energy of the light from the monochromator. Pneumatic detectors are satisfactory over a wide frequency range and, since they respond to total light energy received as distinct from energy received per unit area (thermocouples, bolometers), the light beam from the exit slit need not be focused on to a small area. Thus, some slight simplification of design is possible.

Attenuators—These vary in shape from combs or wedges to rapidly rotating, toothed starwheels. All are designed and precision-manufactured so that the distance moved by the attenuator into or out of the reference beam is linearly related to the increasing, or decreasing, percentage absorption of the sample. Even so, it is impossible to design an attenuator which gives accurate reference beam cut-off close to 0 per cent transmittance. Another disadvantage of an optical attenuator is the loss of balancing information at low transmittance due to the cut-off of light by the comb as well as the sample; this leaves only a weak energy beam for the detector.

Recorders—With the exception of the Zeiss UR-10 spectrophotometer, which gives a needle trace on waxed paper, all instruments record the spectrum as a pen trace on a paper chart with wave number or wavelength plotted versus percentage transmittance, percentage absorption, or absorbance. Movement of the pen is synchronized with attenuator motion while the chart itself is moved at right angles to the pen on a rotating cylinder. A single motor controls both the chart cylinder rotation and the rotation of the Littrow mirror cam and this ensures the identity of the light frequency reaching the exit slit with the frequency position of the pen. By variation of the motor speed, different speeds of scan may be selected.

Sundry figures of infra-red spectra reproduced throughout the text are intended to illustrate also the variation of scales, units and calibration methods employed by manufacturers for their spectral chart paper.

Constant Energy Signals: Slit Programming: 'Gain'

The strength of a detector signal is directly proportional to the 'absolute difference' in energy of the transmitted reference and sample beams. It is vital that at widely differing frequencies the attenuator and pen should receive the same strength signal for identical sample percentage absorptions, otherwise quantitative analysis is impossible. This would be simple if the energy emitted by the source remained constant with changing frequency; this is not the case, however. As shown in *Figure 4*, a large drop in beam energy occurs during the change from 4,000 to 650 cm^{-1}. Two general methods are available to counteract this effect involving, respectively, slit control and electrical 'gain' control.

Slit width may be programmed to increase as the emitted source energy decreases, so that constant reference beam energy enters the monochromator when no sample absorption occurs (no attenuator cut-off of reference beam). Slit programming is installed in most commercially produced double-beam instruments and controlled by a cam of critical design. Mechanical power for driving the cam derives from the same motor which turns the Littrow mirror cam, thus relating the two variables, slit width and frequency. Since resolution of the spectrum is a function of slit width, every spectrum run will have the same resolution at any given frequency, regardless of the sample absorption, providing other possible variables are held constant.

An alternative method of slit control is the constant energy method when the reference beam energy is monitored and the slit servo-operated for constant energy. Thus, servo-slit operation will give increased and decreased slit widths as sample absorption increases or decreases, superimposed upon a steady widening of the slits due to energy decrease with falling frequency. A more erratic slit movement therefore results, especially when a series of sharp absorption bands develops. Cam slit programming does not compensate for the reduced reference beam energy caused by the attenuator movements, whereas monitoring does. However, the monitor and servo scheme has a disadvantage in that resolution at a given frequency will not now be constant for spectra of different compounds.

A second method of obtaining constant pen-recorded percentage transmittance values for identical sample percentage absorption at any frequency is to boost electrically ('gain') the signal from the detector. Gain may be programmed by a cam or servo-controlled, in a similar manner to slit width variation, to ensure that the signal energy reaching the attenuator and pen is no longer a function of frequency. Unfortunately this method usually results in an unacceptably large noise level* in the instrument. Standard practice is to employ a cam for slit programming and a fixed gain. Manual operation of both gain and slit-width is offered as an alternative on most precision spectrophotometers, but the organic chemist may need neither for routine work.

Factors Determining the Quality of a Spectrum

Many inter-related parameters exist in the modern infra-red spectrophotometer; variation of any one of these will affect most of the others. The influences of solvent, phase, prisms and gratings are considered in later sections, while a critical dependence of spectral resolution on slit width has been remarked upon above. For routine laboratory spectral analysis the organic chemist will use either a low-cost instrument, where the manufacturer has already chosen a compromise set of parameters compatible with an acceptable spectrum, or a precision spectrophotometer. With the latter he is more likely to vary the 'rate of scan' among the controls available to him. Every recording pen requires a finite time to respond fully to a signal from the detector. The time required for total deflection of the pen from 100 to 0 per cent transmittance is a variable but, even at its fastest, is seldom less than 2 sec. Hence the definition of a spectrum, the resolution of the bands, will be dependent upon the rate of scan. Slow rates of scan and narrow slit widths are necessary for best resolution (see *Figure 5* and *Table 2*). Furthermore, a slow pen response gives a band outline more accurately traced for intensity measurements. Several instruments incorporate automatic suppression, the extent controllable by the operator, whereby the machine scans rapidly in regions of low absorption but automatically suppresses the rate of scan when a band develops. This important development permits a short overall scan time coupled with high resolution.

Scale expansions of both ordinate (percentage transmittance) and abscissa

* 'Noise' is the appearance of random signals in the recorded spectrum which do not correspond to variations in energy of transmitted light. The extra signals, usually small, arise from numerous points from the detector onwards.

(wave number) spread out the spectrum and increase the separation of the bands. Ordinate and abscissa expansions are equivalent to reduced pen response time and reduced rate of scan, respectively. Many possible combinations of the two variables, rate of scan and wave number scale, exist and

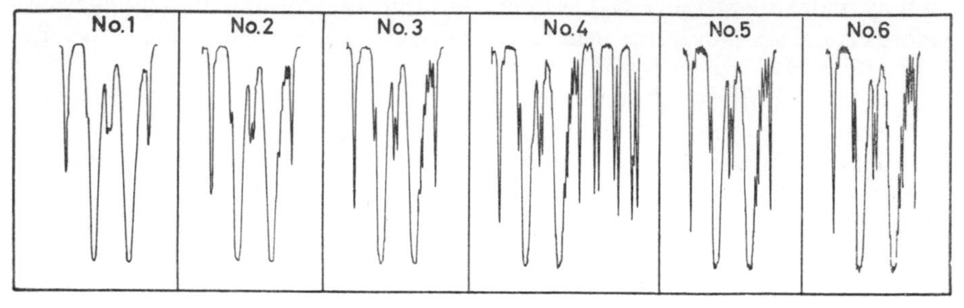

Figure 5. Six spectra of ammonia gas run on a Hitachi Model EP1-2 spectrophotometer with varying slit width and speed of scan (*Table 2*) over the range of 900–1000 cm^{-1} (10·00–11·11 μ). (*By courtesy of* Nissei-Sangyo Co.)

Table 2. Operating Conditions for Spectra Shown in Figure 5

Gain	Constant					
Prism	NaCl					
Abscissa scale	8 cm/100 cm^{-1}					
Spectrum number	1	2	3	4	5	6
Slit width (mm)	0·3	0·2	0·15	0·13	0·12	0·1
Scan time (min) 100 cm^{-1}	0·75	1·5	3	5	10	20

provide valuable flexibility in operating time per spectrum. Collected data on resolution, speeds of scan, chart scales and linearities, together with much information on instrument components, is included in *Table 3*, which covers most commercial double-beam infra-red spectrophotometers in production in June 1963.

Atmospheric fluctuations can cause inconsistencies in instrument performance, principally through deterioration of moisture-sensitive optical surfaces owing to excessive humidity, and the variation with temperature of the refractive index of the prism and hence of its dispersion. Furthermore, atmospheric water vapour and carbon dioxide will cause undesirable energy losses at frequencies where these gases absorb. Strict control of these factors is implemented by one or a combination of several methods, among which are total air-conditioning of the spectrophotometer room, thermostatting of the monochromator above room temperature, liberal use of desiccants and optical systems under vacuum. Where desiccants are used instrument size must be considered, since a large-volume machine with ill-fitting windows and casings will require frequent replenishment. Unduly large dimensions are, in any case, inconvenient for installation of the machine. Removal of water vapour and carbon dioxide by air-conditioning eliminates their absorption bands from the recorded spectrum.

Wave number accuracy and reproducibility* are dependent upon the refractive index of the prism and hence on a constant prism temperature. Some instruments compensate for changes in temperature by altering the optics with a bimetal strip. Prisms may be thermostatted at monochromator temperature during storage or, after prism interchange, a period of several hours is allowed to elapse for temperature equilibration before proceeding with the next spectrum. Even so, calibration of each spectrum against a standard is a recommended procedure. Polystyrene is commonly employed for this purpose (p. 42), with ammonia gas and water vapour being used less frequently. Accuracy and reproducibility of percentage transmittance values are not of immediate importance for routine work in qualitative analysis, and no standard compound has yet been widely accepted for their calibration. Many factors are involved, but most instruments in routine use afford acceptable percentage transmittance values.

Records may be made on two types of printed chart papers—pre-calibrated or instrument-calibrated. Though the former allow easy reading of absorption band frequencies, the wave number accuracy and reproducibility rest on the ability of the operator to place the pen accurately on the chart at the identical frequency with that shown by the wave number counter, on printing accuracy and on the stability of the paper in storage (e.g. no stretching). Furthermore, mechanical connection of the wave number counter with the drum drive is through a common motor and a series of gears. In the course of returning from low to high frequency, after running a spectrum, the 'play' in these gears is taken up in the reverse direction. A residual backlash in the gears must therefore be taken up by running the machine forward before the pen position on the chart is again synchronized with the wave number counter. Unless this procedure is carried out, a constant error of several cm^{-1} will ensue for spectra run on pre-calibrated charts, since the drum plus paper will rotate fractionally before the wave number counter is driven by forward scan. With self-calibrated charts the instrument places its first wave number mark accurately after uptake of the backlash (Unicam's Model SP 100 possesses a built-in correction for this factor). However, since a calibration at two or three widely spaced frequencies with a polystyrene film should occupy less than two min, pre-calibrated charts find favour for routine work. To avoid compression of absorption bands at frequencies below 2,000 cm^{-1} spectrophotometers linear in wave number are equipped for scale expansion at this frequency, customarily by a ratio of 4:1 or 5:1. Linear wavelength scales do not suffer this disadvantage and are, in consequence, the common choice for low-cost instruments to avoid the extra mechanical complexity. Whether automatic or manual change is chosen, scale change exactly at 2,000 cm^{-1} is vital to the wave number accuracy and reproducibility of the spectrum.

There is a critical relation between the frequency shown by the wave number counter and the position of the Littrow mirror relative to the prism, since this position governs the frequency of monochromatic light which passes through the exit slit. For constant wave number accuracy it is therefore of paramount importance that a prism be most accurately placed on its mounting with respect to the Littrow mirror. This source of error is combated

* Accuracy = agreement with the accepted 'true' value. Reproducibility = agreement with measurements previously made on the same instrument.

by the manufacturer with precision-machining of prism mounts and accurate setting up of the optics.

Operators' Requirements

Certain properties of an infra-red spectrophotometer are demanded by all operators, whether their interests lie in organic, physical, theoretical or analytical chemistry. Among these may be included operational reliability, ease of maintenance, good supplies of spares and after-sales service and quick location of faults. Given these primary assets the requirements of the organic chemist then diverge from the person with more theoretical interests. The latter, particularly the research spectroscopist, places a premium on quality of the spectrum. For the highest instrument performance general air-conditioning, reduction of light-scattering at many points and a strict control of all variables receive careful attention. These specifications led to the development of precision spectrophotometers; collected data on their construction and operational limits appear in *Table 3*.

Though precision instruments are extensively used in organic chemistry, their accuracy limits are in excess of requirements for routine work while the profusion of controls is an embarrassment to the beginner. Fine resolution, high transmittance and wave number accuracy plus reproducibility are all welcome but only providing they are a corollary of a rapid rate of scan. Routine work necessitates quick results, i.e. rapidly-run spectra, high instrument utilization and easily read and compared spectra. Simple, robust, inexpensive double-beam spectrophotometers have been developed to meet this demand and are now the 'hack' of the general research and teaching laboratories. A compromise of the various parameters is selected to reduce operating controls to a mere five or six. Incorporation of a single sodium chloride prism suffices for most needs. A recent survey[20] revealed that 75 per cent of low-cost instruments were purchased for organic and analytical chemists, many of them in Universities and, of the spectra run by these two groups, 85 per cent were for qualitative analytical purposes*. The first part of *Table 3* summarizes details of the construction and performance of the available instruments.

Individual Instruments†

Double-beam infra-red spectrophotometers are currently manufactured in six countries. All demonstrate certain common features (*Table 3*) while some possess novel devices. Accessories available with the best precision instruments include polarizers, microscope attachments, multi-pass gas cells, a range of cells for liquid phase spectra, presses and dies for alkali

---

* In the Organic Research Laboratories at Imperial College a Perkin-Elmer 137 was used by approximately 40 different operators annually with a turnover of *ca.* 4,000 spectra, virtually all in routine qualitative analytical work.

† A survey of world sales would reveal that certain models are internationally distributed while others are confined within regional or national boundaries. Numbers sold should not be considered an ultimate criterion of technical superiority. Many other factors are involved, some bearing little relation to instrumental performance. Detailed comparisons show no single model to be pronouncedly superior to its competitors, and claims to the contrary should be viewed with reserve. Opinions expressed here are those developed from personal experience and many discussions with operators from several nations.

halide disc preparation, and a selection of prisms and gratings. These are considered in later sections.

Several collections of reference spectra (see p. 46) have been progressively developed and the infra-red spectra of several thousands of compounds have already been documented, many on card index systems. Several instruments are adaptable for recording on to punch cards or small file charts either directly or via slave recorders, which should materially assist the expansion of these infra-red spectra collections. All users of infra-red spectroscopy should benefit from the increasing availability of low-cost instruments with monochromators equipped with gratings rather than prisms. The former offer an improved signal : noise ratio leading to a superior resolution at higher recording speed.

CELLS AND SAMPLING TECHNIQUES

INFRA-RED spectra of both gases and liquids may be obtained by direct study of the undiluted specimen; solids, however, are usually studied after dispersion in one of a number of possible media. Various cells and standard sampling procedures are outlined here for each phase.

A general warning on the use of solvents is appropriate at this point. The toxicity of solvents commonly employed for solution spectra and the cleansing of cell plates should be guarded against; carbon disulphide and chlorinated hydrocarbons are particularly noxious. Air-conditioning of a closed room may involve no intake of fresh air but only a cycling of the same air with continuous removal of water vapour and carbon dioxide. Therefore, solvent vapours may steadily accumulate to a concentration deleterious to both instrument and operator. The vapour in contact with the high-temperature radiation source may be pyrolysed and yield corrosive gases; chloride corrosion deposits have been observed on mirror mounts and the casing surrounding the source filament. Periodic or slow continuous flushing with dry air or nitrogen will preserve the instrument. Since the same remedy is denied the operator, sample preparation and cell-cleansing procedure should always be performed in a well-ventilated laboratory or fume chamber away from the spectrophotometer. Nitrogen flushing of the instrument should give a gas flow through the monochromator to the source chamber, and then out of the instrument. Reversal of this flow direction can lead to corrosion within the monochromator by nitrogen oxides, formed from nitrogen and air at the high-temperature surface of the source.

Crystalline Solids

Three general methods are available for the examination of solids in their crystalline form. All involve the reduction of a solid to very small particles, which are then diluted in a mull, in an alkali halide disc, or spread as pure solid on a cell plate surface. Though, ideally, complete breakdown of crystals into individual unit cells is required for the best solid-phase spectra, this degree of particle division is never attained. Even in alkali halide discs, where particle size reduction is most efficient, the particles are still microcrystalline. Electron microscope examination has revealed that discs contain several unit cells per solid particle. Both the orientation of molecules with respect to each other and intermolecular interactions complicate the infra-red spectrum.

Mulls—For mulls, Nujol (high-boiling fractions from petroleum) is the most commonly used mulling agent. Fluorolube (perfluorokerosene, a mixture of fluorinated hydrocarbons) and hexachlorobutadiene have both been employed as mulling agents when it was desired to study frequency ranges in which Nujol absorption bands appear. The fluorinated hydrocarbon mixture, however, is difficult to remove from cell plate surfaces and sample recovery is troublesome.

Making good mulls is an art, and experience is the best tutor. The sample (2–5 mg) is well ground together with the mulling agent (1 drop), either in a ball-mill or simply with a pestle and mortar. If the latter method is employed great care must be exercised to ensure that the larger aggregates of solid, which tend to move away from the region covered by the circular rubbing motion of the pestle, are continually brought back to the centre of the mortar to be crushed. Substances which mull easily are ground to a satisfactory

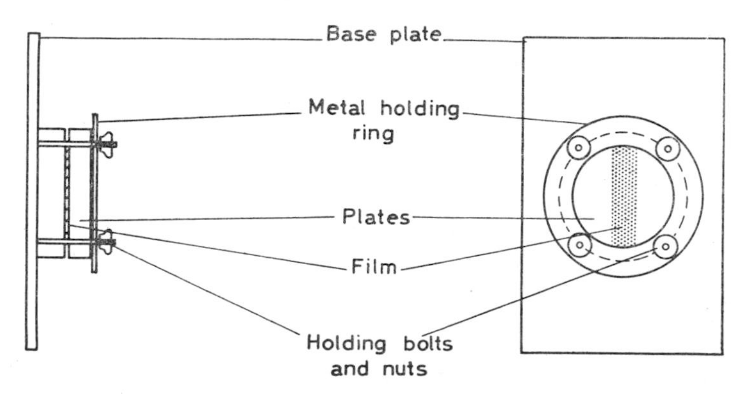

Figure 6. Typical plate assembly and holder

state in 1–2 min. Less tractable materials may require up to 30 min of grinding before a useful mull is obtained. When the grinding is complete the mull is transferred to a rock-salt plate by means of a fine spatula or, better, with a razor blade. The mull is covered by a second plate which is pressed lightly to force the mull to spread as a thin film. Gentle rotatory manipulation of the plates can be carried out to alter the thickness of the film or to attain a more even distribution. The plates are then placed in the sample beam path of the infra-red spectrophotometer, being retained there in a cell plate holder (e.g. *Figure 6*) or in a simple push-fit holder. Next, a few fast runs are made of a region of the spectrum where absorption appears to be strong—e.g. 2000–1500 cm^{-1} (5·0–6·7 μ)—to ascertain that the thickness of the mull film is suitable for the obtaining of a useful spectrum. After any of these preliminary runs the assembled plates plus sample can be removed from the holder and the film thickness altered, as described above, until a satisfactory trial spectrum results. Care must be taken that the absorption peaks used as a check of the suitability of the film thickness are those of solute and not of the mulling agent. Absorption frequencies of the latter are listed elsewhere (*Table 5*, p. 29). Lithium fluoride, calcium fluoride and potassium bromide are alternative materials for the plates, the latter varying in size according to the instrument in use and usually being up to 0·5 cm thick, 1–3 cm diam., and always brittle. Since the infra-red light incident upon the plates is in the form of a thin beam, there is no real necessity for film dimensions in excess of 0·5 × 1–2 cm; plates are of slightly larger dimensions and may be circular to facilitate manufacture. It is not generally possible to obtain reliable intensity values from mull spectra since there is no simple control of sample thickness or of concentration (see, however, p. 37).

In view of the ease with which cell plates may be damaged, a few remarks on handling techniques are pertinent. Alkali halide plates are attacked by moisture and they must therefore be stored in desiccators, or at *ca.* 40°, and handled appropriately; breath, or moisture from operators' hands, can mar the plate surfaces. Because the sample is in microparticle form losses of radiation occur by scattering. Similar losses occur from non-planar rough plate surfaces and, for best results, these should be polished before each spectrum is taken. In routine laboratory work this procedure is often carried out at less frequent intervals. Cleansing after use is accomplished simply by brief immersion of the plates in chloroform or other dry solvent, and quick drying with lens tissues or other dry, soft material. Solvent should not be allowed to evaporate from the plate surfaces, otherwise they are cooled and traces of moisture condense upon them.

In practice it will be observed that all alkali halide plates give a constant background absorption. This may be corrected best by adjustment of the 100 per cent transmittance control, or by insertion of a single plate in the reference beam of the instrument. Comparison of spectra recorded with well-polished and unpolished plates reveals the losses due to light-scattering for unpolished surfaces. Polishing is effected manually by taking the plate, moistening the flat surfaces with ethanol, and then rubbing with a soft chamois leather previously impregnated with rouge. Rubbing is carried out with circular motions of the finger tips, changing the grip on the plate frequently. In such a manner it is possible to smooth the fogged surfaces. Transferring the plate to a dry portion of the chamois leather, rubbing with circular motions is continued until the plate is dry and well-polished. In practice care has to be taken to avoid rubbing the centre of the plate more than the outer parts; otherwise, the plate surface soon becomes concave, rather than flat, and thin films can no longer be obtained by pressing two plates together.

Solid deposits—Samples for solid state spectra may also be obtained by deposition of the solid on a plate surface from a solution. A concentrated solution is allowed to evaporate slowly on the plate surface, leaving an even layer of solid as a glassy film. A large number of solvents or mixtures of solvents may have to be experimented with before the correct form of deposit is obtained. Layers of powdery crystals or scattered large crystals give poor spectra because of heavy radiation losses by scattering. Low-melting solids may be melted between two plates, the melt pressed into a film, and the latter allowed to solidify by slow cooling. Solidified melts often give trouble due to uniform orientation of crystals. Solid state spectral study by deposition of samples is not a common practice.

Pressed discs—Pure, dry alkali halide (150–250 mg)—potassium chloride or potassium bromide—is intimately ground together with the solid sample as described in mull preparation. Suggested concentrations are 1 mg solid per 100 mg alkali halide for a substance with molecular weight of 200, and proportionately more solid with increasing molecular weight. Compression of the mixture at room temperature, under vacuum and high pressure, furnishes a solid disc, usually transparent. Pressures of up to 100,000 lb/in^2 have been recommended, but most manufacturers suggest pressures in the range 20,000–50,000 lb/in^2. The disc is mounted in a holder, fitted with a metal well the shape of the disc, and held in the sample beam path. Discs have an advantage over mulls in that the concentration and thickness of the specimen is readily determined, thus permitting their use in quantitative analyses (p. 37). However, scattering losses are again incurred, and homogeneous dispersal of the sample is not always easy. Evaluation of these losses is difficult and their reproducibility questionable. Simple laboratory hand presses for disc preparation are marketed by most instrument manufacturers, though hydraulic presses are more convenient for routine use. Dies must be scrupulously cleaned after use to remove all traces of alkali halide which corrodes stainless steels. Anomalous spectra due to physical or chemical changes induced by grinding have been reported in the chemistry of carbohydrates[21], penicillin[22], polyhydroxysteroids[23] and some simpler molecules[24].

Liquid-phase Spectra

Pure liquids, and solutions of solids, liquids or gases may be examined in the infra-red. Liquids are generally not examined in solution, thus avoiding solvent absorption interference, though a solution spectrum may remove some of the intermolecular interactions existent in the pure liquid. Solutions may also be necessary when sample absorption is strong and insufficiently thin films are readily and accurately available.

Pure liquids—The description for practical work with mulls also applies to thin liquid films. However, since intensity measurements and spectral comparisons are more readily made when sample thickness is controllable, cells of the type employed for solution spectra may be preferred, with path lengths of 0·005–0·1 mm. Compensation is necessary for cell plate background absorption, as outlined above for solid-state spectra. Volatile liquids which evaporate from inter-plate films must be examined in a sealed cell.

Table 4. Solution Spectra Requirements—the Relation between Path Length, Solution Volume, Sample Weight, and Molecular Weight

Cell path length (mm)	Molecular weight							Volume of solvent required (ml.)
	100	150	200	250	300	400	500	
	Weights of solute required (mg)							
0·2	3·5	5	6	8	10	13	16	0·2
0·5	1·7	2·5	3	4	5	6·5	8	0·25
1	1·7	2·5	3	4	5	6·5	8	0·5

Spectra of solutions—Chloroform, carbon tetrachloride and carbon disulphide are the three solvents most commonly employed, in cells of 0·1 mm–1 cm thickness. Concentration effects may be studied using the same thickness cell, and the intensities of both weak and strong absorption bands can be determined at one concentration by use of thick and thin cells, respectively. *Table 4* gives, for three different standard cells, the volumes of solvent and weights of solute, dependent on molecular weight, recommended for production of average spectra.

It should be noted that none of these standard cells should have a capacity in excess of 0·5 ml.; if it has, then the design is at fault. Accurate intensity measurements are possible when the molar concentration of solute is known. Careful transfer of the prepared solution to the cell by means of a pipette or hypodermic syringe reduces evaporation and transfer losses, so that 0·25 ml. solution should be ample volume for a 0·5 mm cell. Where emphasis is placed on intensity measurements many operators prefer to prepare a larger volume (1–2 ml.) of solution and thus reduce errors in sample preparation and manipulation.

Solution cells are of two types, i.e. demountable cells, the separate components of which are assembled by the operator for each spectrum, and fully assembled sealed cells of fixed cell path length, each type possessing its merits and disadvantages. Demountable cells are more easily cleaned, and the plate surfaces may be polished prior to each spectrum. Complete, sealed cells are cleaned by repeated flushing with solvent, but when the plate surfaces deteriorate, dismantling and overhaul can prove expensive and time-consuming. Furthermore, for each pair of demountable cells a whole range of metal, or solvent-resistant plastic, spacers may be utilized to give a series of cells of varying path lengths using various plate materials. This flexibility is not available with sealed cells where duplication of plates and holders is necessary for each cell thickness. However, in the average laboratory most of the solution spectra may be obtained with the same path length (*ca.* 0·5 mm), and a set of five pairs with path lengths of 0·05, 0·1, 0·5, 1·0 and 2·0 mm will suffice in all but exceptional cases. Whenever solution cells are used, a compensating cell containing an equal thickness of pure solvent is necessary in the reference beam. The recorded spectrum is then that of the solution minus the solvent and, providing the path length through each is closely similar, this will be the spectrum of the solute, except in those regions where the solvent absorbs strongly. For this condition to be met, cells of almost equal path length must be available, and the current procedure is to have matched cells, with carefully machined spacers to ensure identical thicknesses. On this account the fixed, complete cells are superior since, in these, the spacers are not subject to damage by handling, which does happen when demountable cells are repeatedly assembled and dismantled. An alternative to matched solution cells is a variable-thickness, compensating solvent cell which has a micrometer device attached to one movable plate, thus allowing selection of any cell thickness. Each solution cell may then be 'calibrated', and the correct setting of the variable cell thickness predetermined for compensation of solvent absorption. These variable-thickness cells are expensive. Solution cell designs (e.g. *Figure 7*) resemble cells for liquid films (*Figure 6*) with addition of a spacer between the two cell plates. Narrow inlet and outlet ports are drilled through both metal and alkali halide top plates for injection of solution into the cavity by means of a small pipette or hypodermic needle and to permit the escape of trapped air bubbles. Plastic stoppers keep the solution within the cell, and the specimen is recoverable in good yield after the whole operation has been completed.

Precautions should be taken to exclude all silicones from sealed cells, since they adhere to the plate surfaces and give rise to anomalous, strong, broad absorption at 1,100–1,000 cm^{-1} (9–10 μ), thus obliterating a useful region.

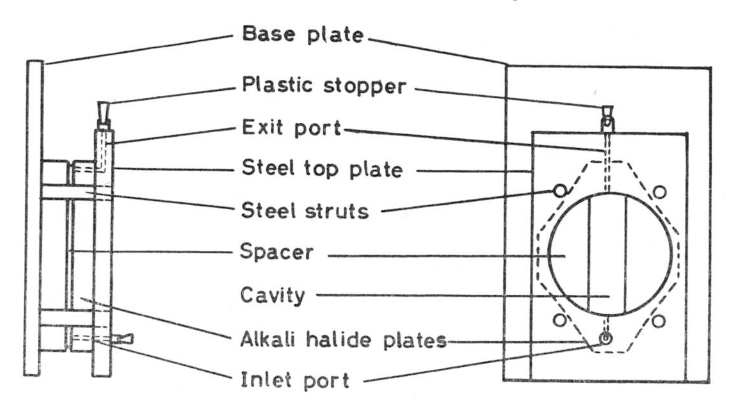

Figure 7. Typical sealed solution cell

Gas–liquid chromatography columns may have a silicone oil phase, and several silicone, high-vacuum, stopcock greases exist; these are two potential sources of contaminant.

Caution must be exercised in the interpretation of infra-red spectra obtained for solutions of compounds capable of intermolecular association. In these cases bands may appear at different frequencies for associated and non-associated molecules over quite a wide range of concentration. Such phenomena are not to be confused with absorptions due to impurities, or with those cases where two absorptions result from in-phase and out-of-phase vibrations. Intermolecular associations are considered in greater detail on p. 40.

Specific problems concerning the measurement of infra-red spectra using aqueous solutions are discussed in the next short chapter (p. 31).

Spectra of Gases

Gas-phase infra-red spectra differ fundamentally from condensed-phase spectra, principally because the molecules are free to rotate in a gas and intermolecular interaction is minimal. This gives rise, in simple molecules, to an abundance of fine structure, corresponding to rotational energy level transitions (*Figure 8*). For this reason, various gases are valuable for frequency calibration; ammonia, carbon dioxide and water vapour are examples. In view of their reduced molecular concentrations, gas cell path lengths are desirable in centimetres rather than in fractions of millimetres used for condensed phases. Shorter path lengths for gases are possible when high pressure gas cells are employed.

Special techniques have been developed for weighing small quantities of gas and effecting total transference to the gas cell. In a common procedure, however, the quantity of gas admitted to a standard volume cell is known from measurement of its pressure. To fill the cell it is first evacuated and the gas is then passed in along the pressure gradient.

Available straight-line paths between the source and monochromator entrance slits seldom exceed 25 cm (the Hilger-Watts Model H-800 is exceptional in allowing a 50 cm straight-path gas cell). For low pressures, or

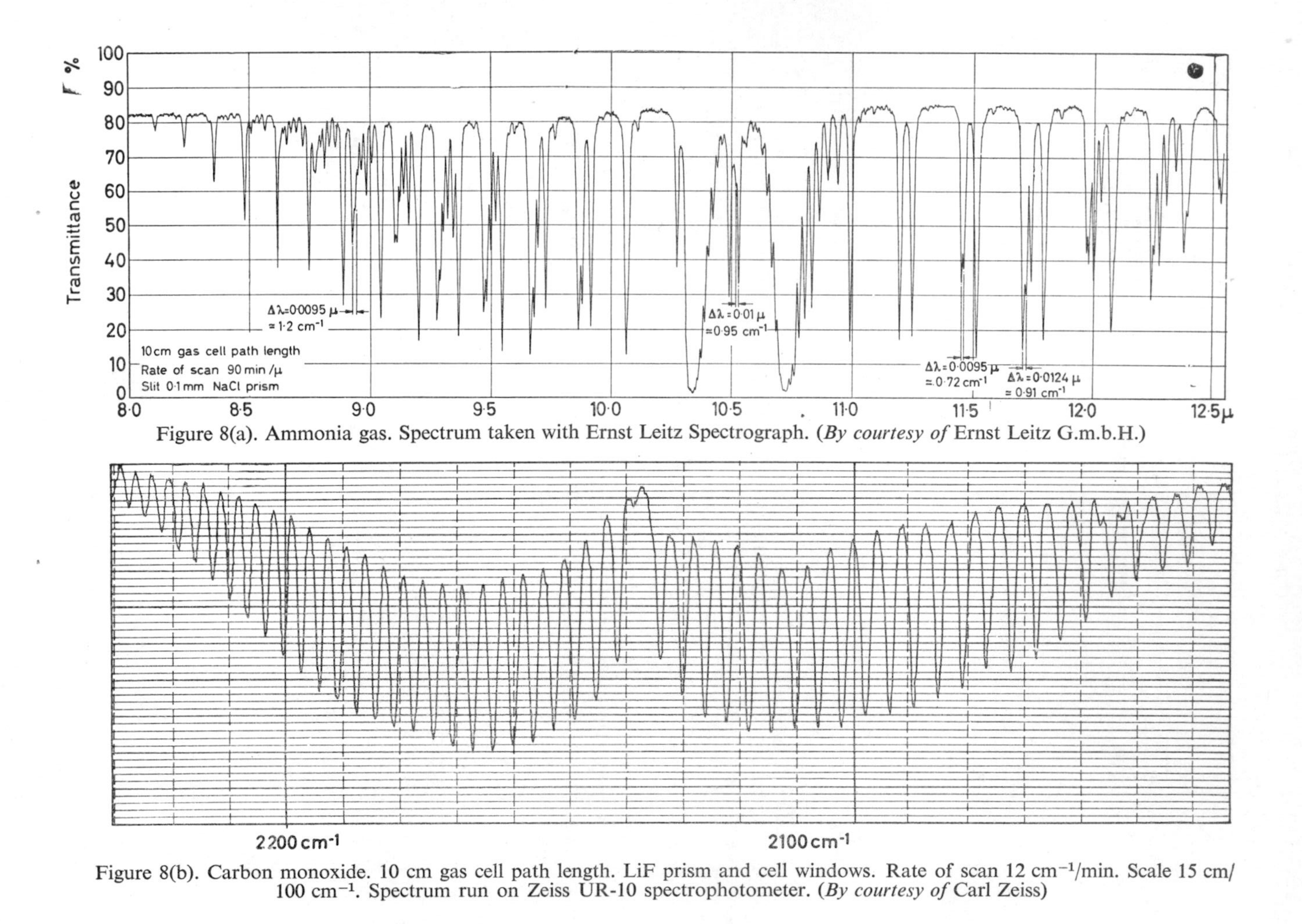

Figure 8(a). Ammonia gas. Spectrum taken with Ernst Leitz Spectrograph. (*By courtesy of* Ernst Leitz G.m.b.H.)

Figure 8(b). Carbon monoxide. 10 cm gas cell path length. LiF prism and cell windows. Rate of scan 12 cm^{-1}/min. Scale 15 cm/100 cm^{-1}. Spectrum run on Zeiss UR-10 spectrophotometer. (*By courtesy of* Carl Zeiss)

mixtures of gases, and particularly for analysis of trace quantities of gas in the atmosphere, special long-path gas cells have been developed. The light beam is deflected through 90° and repeatedly reflected between the ends of a long (up to 4 m) gas cell by judicious use of mirrors, before being turned again through 90° and into the monochromator (*Figure 9*).

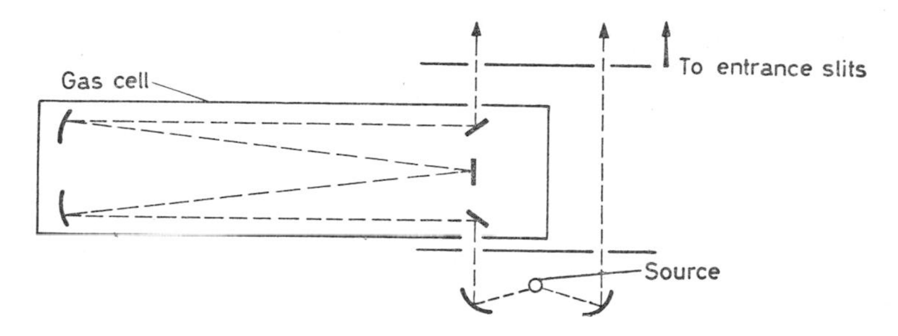

Figure 9. Simplified typical multi-reflection gas cell

Microscopy Infra-red Study

In recent years commercial microscope attachments have appeared for use with standard instruments; these focus the light beam from the source on to a minute area and accurate location of the microsample (single crystal, fibre, etc.) at this point allows the whole light beam to pass through a minimum of sample. The transmitted beam is then magnified before reaching the monochromator entrance slit. Such micro-methods present considerable practical difficulties in solution manipulation and transfer to micro-cells.

As an example of the newer devices available there may be quoted the micro-sampling accessory to several Perkin-Elmer Models. Here the infra-red beam is reduced to 1/36th its normal area. An ultra-micro KBr die producing KBr discs as small as 0·5 mm in diameter is also available.

THE CORRECT PHASE AND SAMPLE DILUENT

A SINGLE infra-red spectrum will not provide all the valuable information which it is possible to gain by infra-red spectroscopic study of a compound; this statement applies to all phases and all sample diluents. Change of phase or diluent will frequently yield additional evidence, even if this only confirms conclusions drawn from the first spectrum. However, one spectral region is often of greater intrinsic value than others, and a wise choice of sampling procedure can be most beneficial; some substances are best examined in one particular phase. In this section the effects of phase and diluent changes, the absorptions of the diluents themselves, and some of the pitfalls and advantages of specific diluents will be discussed.

Selection of Phase

There are no rigid rules concerning the phase to be preferred for any type of compound, but pure liquids or dilute solution spectra are best for general use, unless there are good practical or theoretical reasons for choosing otherwise. Dilute solution spectra with non-polar solvents largely eliminate solute intermolecular interactions, though some major interactions may persist (see Hydrogen Bonding, p. 40). Unfortunately, new limitations are introduced in the form of solubilities and solvent absorption. Furthermore, if from solubility considerations polar solvents are necessary, then solute–solvent interactions develop. No single solvent is satisfactory for the whole range 4,000–650 cm^{-1}, and study of solutions in two or more different solvents may be expedient for unknown substances. Major practical advantages of solution spectra include ease of preparation, uniformity of dispersion of solute and ready control of both concentration and path length.

Carbonyl-containing compounds are studied in dilute solutions and much valuable information has accumulated on both absorption band position and intensity; non-polar solvents are preferred. Carboxylic acids, normally existent in dimeric form owing to association, are investigated in the solid or pure liquid phase where they are fully associated. In solution their degree of association is highly dependent on the polarity of the solvent; in non-polar solvents the acids remain in dimeric form but in the most polar solvents solute association is replaced by a solute–solvent interaction. Amides also exhibit carbonyl C=O stretching absorption, but solute association is weaker because hydrogen bonding of the amino group is weaker than that of hydroxyl. Association of amides therefore shows greater sensitivity towards weakly polar solvents than does carboxyl association.

Absorption by the Diluent

Solid-phase spectra—Alkali halides for discs must be dried to avoid weak, broad, hydroxyl background absorption in the region of 3,700 cm^{-1} (2·70 μ) and 1,600 cm^{-1} (6·25 μ). Nujol (*Figure 10*), Fluorolube and hexachlorobutadiene give a few strong bands, and interpretation of bands in these narrow-frequency ranges is therefore meaningless. Data for various sample diluents absorptions is given in *Table 5*.

Liquid-phase spectra—The frequency ranges available for spectral examination of any solute in a given solvent are limited by solvent absorptions, but with matched solution and reference (solvent) cells only the major solvent absorption bands interfere. Absorption spectra of four common solvents, at two cell thicknesses, are given in *Figures 11–14*, and the ranges for which

Table 5. Strong Absorptions of Solvents (0·05 mm Thickness) and Oils (Thin Films) in the Range 4,000–650 cm^{-1}

Solvent	Absorption > 25%	
	Frequency (cm^{-1})	Wavelength (μ)
Chloroform	3,010–2,990	3·32– 3·34
	1,240–1,200	8·07– 8·33
	815– 650	12·27–15·38
Carbon tetrachloride	820– 725	12·20–13·79
Carbon disulphide	2,220–2,120	4·50– 4·72
	1,630–1,420	6·14– 7·04
Methylene dichloride	1,290–1,240	7·75– 8·07
	905– 890	11·05–11·24
	800– 600	12·50–16·67
*cyclo*Hexane	3,050–2,800	3·28– 3·57
	1,470–1,420	6·80– 7·04
	1,260–1,250	7·94– 8·00
	915– 850	10·93–11·76
Benzene	3,100–3,000	3·23– 3·33
	1,970–1,950	5·08– 5·13
	1,820–1,800	5·50– 5·56
	1,510–1,450	6·62– 6·90
	1,050–1,020	9·52– 9·80
	700– 650	14·29–15·38
Dioxane	3,100–2,650	3·23– 3·77
	1,490–1,020	6·71– 9·80
	915– 830	10·93–12·05
Bromoform	3,080–3,020	3·25– 3·31
	1,170–1,120	8·55– 8·93

Solvent	Absorption > 25%	
	Frequency (cm^{-1})	Wavelength (μ)
Acetone	3,100–2,880	3·23– 3·47
	1,820–1,170	5·50– 8·55
	1,110–1,070	9·01– 9·35
	915– 885	10·93–11·30
Pyridine	3,200–3,000	3·13– 3·33
	1,630–1,410	6·14– 7.09
	1,230– 970	8·13–10·31
	780– 650	12·82–15·38
Methanol	3,800–2,800	2·63– 3·57
	1,520–1,360	6·58– 7·35
	1,160– 960	8·62–10·42
	710– 650	14·08–15·38
Water*	3,650–2,930	2·74– 3·41
	1,750–1,580	5·71– 6·33
	930– 650	10·75–15·38
Deuterium oxide*	2,780–2,200	3·60– 4·55
	1,280–1,160	7·81– 8·62
Oils		
Nujol	2,920–2,710	3·42– 3·69
	1,470–1,410	6·81– 7·08
	1,380–1,350	7·24– 7·41
Hexachloro-butadiene	1,640–1,510	6·10– 6·60
	1,200–1,140	8·35– 8·77
	1,010– 760	9·90–13·10

* Values for 0·01 mm cells [25]

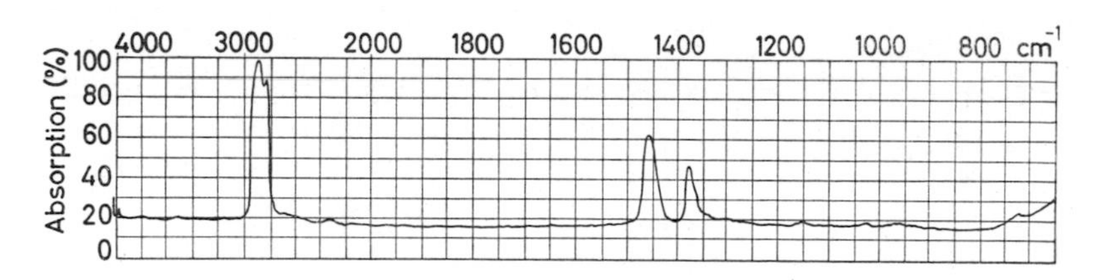

Figure 10. Nujol film on NaCl plate (NaCl prism). Run on Hilger H-800 spectrophotometer (*By courtesy of* Hilger and Watts Ltd.)

interpretations are meaningless are listed for these and several more solvents in *Table 5*. When the cells are distorted from matched status, or when the two beams from the source are out of balance, or when solute concentration is high (finite mole fraction of solute), peaks or troughs will appear which are

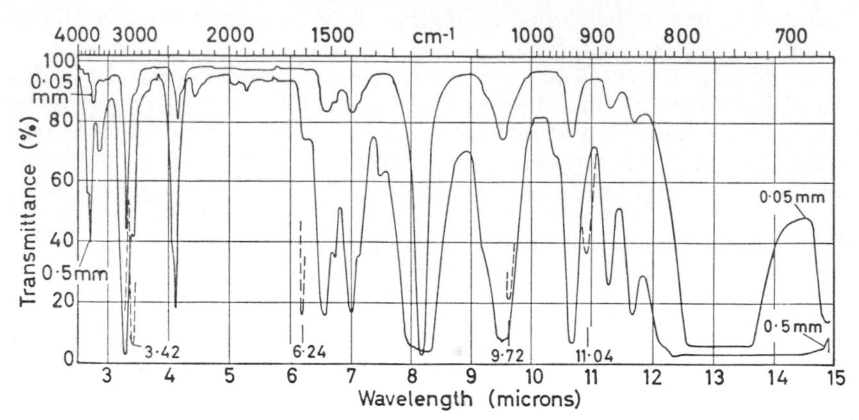

Figure 11. Chloroform. Run on Perkin-Elmer model 137 (Infracord). Each solvent recorded for 0·05 mm and 0·5 mm thickness cells

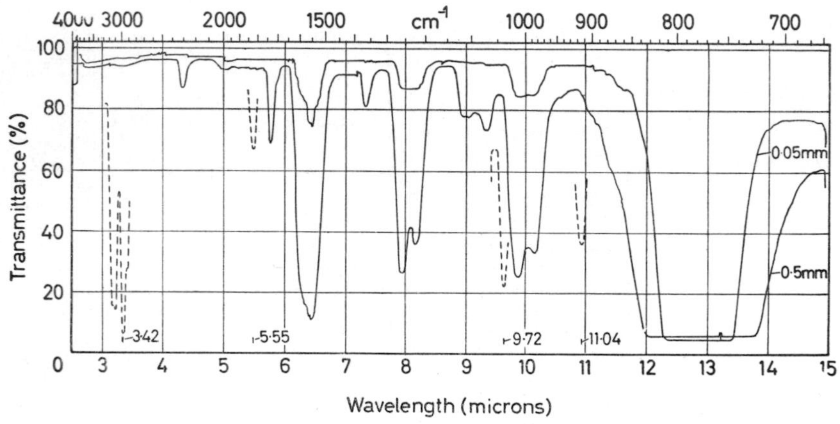

Figure 12. Carbon tetrachloride. Run on Perkin-Elmer model 137 (Infracord). Each solvent recorded for 0·05 mm and 0·5 mm thickness cells

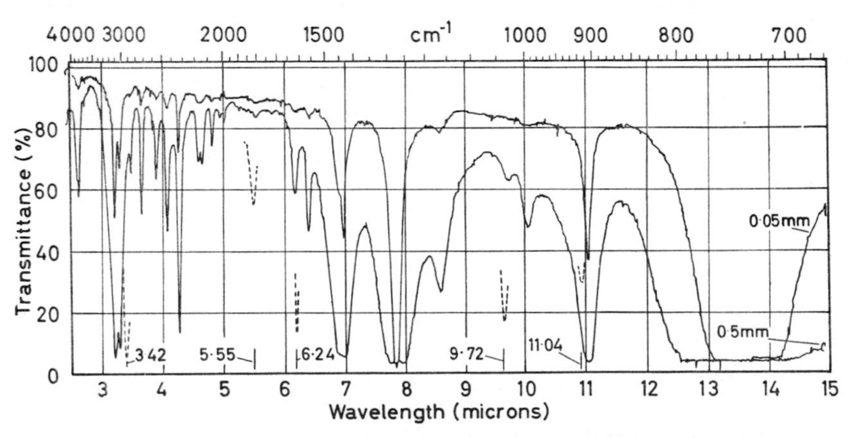

Figure 13. Methylene dichloride. Run on Perkin-Elmer model 137 (Infracord). Each solvent recorded for 0·05 and 0·5 mm thickness cells

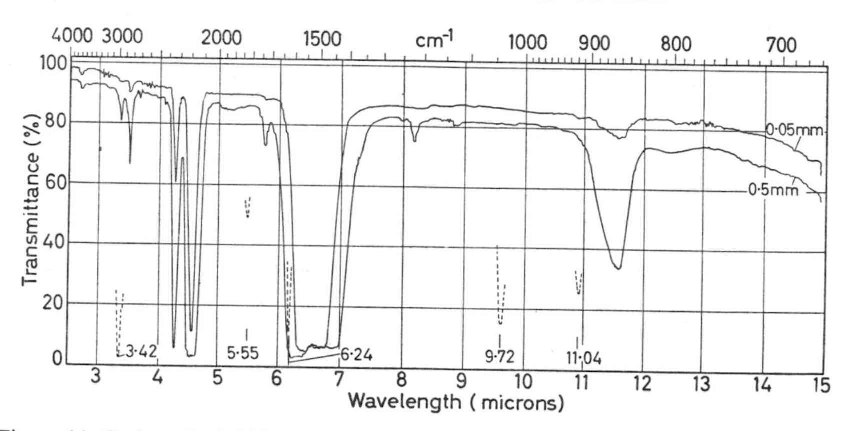

Figure 14. Carbon disulphide. Run on Perkin-Elmer model 137 (Infracord). Each solvent recorded for 0·05 mm and 0·5 mm thickness cells

due to the solvent. Unfortunately, increasing solvent polarity is paralleled by more intense and more frequent solvent absorptions. Organic solvents must always be dry and pure. If desirable, traces of water may be removed from solvents by contact with a zeolite molecular sieve. Commercial chloroform contains ethanol as a preservative which is removable, immediately prior to use, by rapidly passing the solvent through Grade I activated alumina, thus avoiding anomalous hydroxylic absorptions. Primary and secondary amines should not be studied in carbon disulphide solution, since reaction with the solvent may occur to yield alkyldithiocarbamic acids.

The organic chemist will not often turn to water as a solvent for solution spectra. In addition to itself absorbing over broad ranges in the infra-red region, solvent water presents new practical difficulties in the choice of suitable cell window materials. However, several types of windows have been shown to be satisfactory, although each has its disadvantages. Barium and calcium fluoride windows, although transparent down to 940 cm^{-1} (10·5 μ) and 1,250 cm^{-1} (8·0 μ), respectively, are very brittle and acid-sensitive. Arsenic sulfide, As_2S_3, is somewhat more resistant to acids and is transparent down to *ca.* 1,250 cm^{-1} (8·0 μ). Silver chloride windows are apparently unattacked by acids and, moreover, are transparent over the whole infra-red range. They suffer the distinct disadvantage, however, of being light-sensitive. For investigations above 2,200 cm^{-1} (below 4·5 μ) and extending into the near infra-red, cell windows may be constructed of thin glass. Quartz is an alternative for studies above 2,850 cm^{-1} (3·5 μ).

Irrespective of which of these cell window materials is employed, cell path lengths of 0·01 mm or less are necessary owing to the intense absorptions by solvent water. Since the absorptions of deuterium oxide occur over frequency ranges where water is transparent, the infra-red range from 5,000 cm^{-1} to 650 cm^{-1} (2·0–15·4 μ) can be studied by use of both H_2O and D_2O as solvents. It must be noted that deuterium oxide is of value for such a study only when the solutes lack exchangeable hydrogens.

A novel technique employing impervious plastic bags inside regular cells to protect water-sensitive windows has been described[26].

Solvent–Solute Interaction

Solvent association may be a general orientation around the whole solute molecule or about one particular group. Chloroform possesses a polar hydrogen atom and polar chlorine atoms which increase its solvent powers compared with carbon tetrachloride, but do so by virtue of a solvent–solute association. Association causes selective, small-frequency shifts of absorption bands, notably those for C=O, O—H and N—H stretching vibrations, by comparison with spectra for solutions in symmetrical, non-polar carbon disulphide or tetrachloride. Small shifts are observed too for these stretching absorption bands when solutions in benzene, and other solvents containing no polar hydrogen atom, are examined. Acetone absorbs in the range 1,728–1,718 cm^{-1} (5·788–5·822 μ) for C=O stretching, according to the solvent used.

Frequency Shifts by Phase Changes

Absorption band shifts occurring through changes of sample diluent or phase can be of definite benefit in qualitative analysis. Apart from band frequency shifts, band splitting or coalescing may occur, as exemplified below.

N-methyl and N-ethyl acetamide absorb at 1,720–1,715 cm^{-1} (5·81–5·83 μ) as vapours, near 1,700 cm^{-1} (5·88 μ) in dilute solution, and as liquids in the condensed state the frequency falls even further, to 1,650 cm^{-1} (6·06 μ).

A frequency fall of 10–20 cm^{-1} (wavelength increase 0·03–0·07 μ) is usually observed for ketone C = 0 stretching absorption on passing from the vapour phase to pure liquids (the fall may reach 30 cm^{-1} (0·1 μ increase) for esters). Cyclopentanone shows a C = O absorption in the vapour phase at 1,772 cm^{-1} (5·643 μ), whereas a liquid film absorbs at 1,746 and 1,732 cm^{-1} (5·727 and 5·773 μ) (grating instrument). Solutions show the two absorption peaks at 1,741–1,750 cm^{-1} (5·744–5·714 μ) and 1,725–1,732 cm^{-1} (5·798–5·773 μ), frequencies being solvent dependent. Splitting is due to Fermi resonance*[27].

Some primary amides and simple secondary amides, when examined in the solid state, show only a single band at 1,680–1,600 cm^{-1} (5·95–6·25 μ) which is split into the Amide I and Amide II bands in spectra taken with dilute solutions. This particular band cleavage arises from the breaking of intermolecular hydrogen bonds.

Sulphones provide a further example of the band shifts caused by changes of state. The bands appearing at 1,350–1,300 cm^{-1} (7·41–7·69 μ) and 1,160–1,120 cm^{-1} (8·62–8·93 μ) for solution spectra show strong shoulders, and the strong absorption covers a fairly wide frequency range. In the solid state this absorption appears more resolved, as a group of strong bands at closely similar frequencies, but the change to a solid phase spectrum is also accompanied by a shift of *ca.* −20 cm^{-1} (+0·15 μ) in band positions.

Methylene rocking vibration absorption for the $(CH_2)_4$ group at 750–720 cm^{-1} (13·33–13·89 μ) appears as two bands in the crystalline state but only

* Fermi resonance is a coupling between a fundamental band and an overtone or combination band. Coupling occurs when the two bands are at similar frequencies and can result in a substantial increase in intensity of the weaker band and the appearance of a split absorption.

as a singlet with liquids or solutions. Interactions between neighbouring molecules in the crystalline material cause the band duplication; many further examples of the effects of phase changes on spectra are in the original literature. For example, during the structural elucidation of the mould metabolite fumagillin it was found that the carbonyl stretching frequency for the cyclohexanone III was at 1,692 cm^{-1} for a chloroform solution. However, when the spectrum was recorded for a potassium bromide disc this absorption was no longer present, showing that in this medium the compound exists completely in the hemiketal form IV[28].

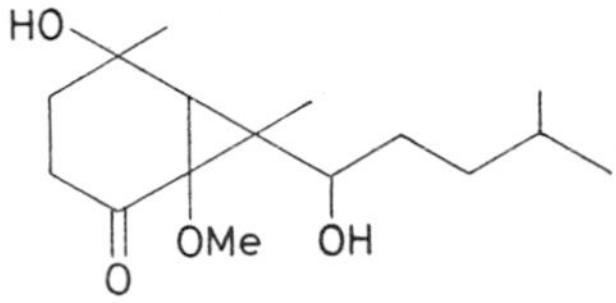

HO OMe OH O

III

O OMe OH OH

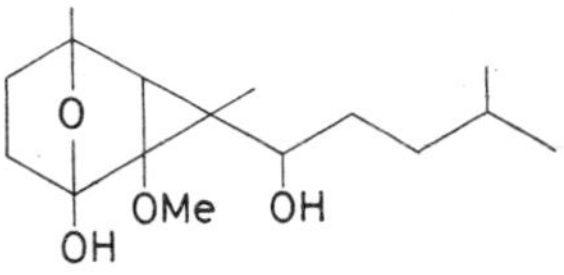

IV

PRISMS AND GRATINGS

THE function of the prism or grating within the monochromator unit of the spectrophotometer has been considered in the section devoted to instruments. Attention was drawn to the unequal dispersion of different prism materials with changing frequency; improvements in spectral resolution may therefore be obtained by correct selection of the prism. *Table 6* gives the useful infra-red frequency ranges for several types of prism, but it is towards the lower frequencies in the quoted ranges that the prism material is most efficient.

Table 6. Prism Frequency Ranges

Prism material	Glass	Quartz	CaF_2	LiF	NaCl	KBr(CsBr)	CsI
Useful frequency range (cm^{-1})	above 3,500	above 2,860	5,000–1,300	5,000–1,700	5,000–650	1,100–285	1,000–200
Wavelength range (μ)	below 2·86	below 3·5	2·0–7·7	2·0–5·9	2–15·4	9–35	10–50

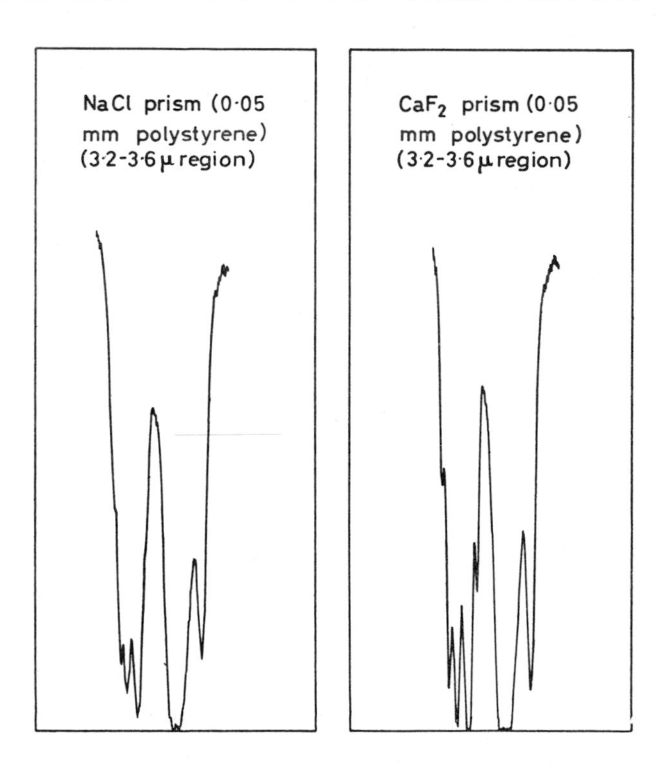

Figure 15. Dependence of resolution at a chosen frequency upon prism material. Both spectra run on a Beckman IR-4 spectrophotometer. (*By courtesy of* Beckman and Co.)

Sodium chloride prisms are chosen as a compromise for the range 4,000–650 cm^{-1}, the fundamental infra-red region, to avoid a multiplicity of prisms and the need for prism interchanges. Rock salt gives acceptable resolution in the important carbonyl stretching range 1,950–1,600 cm^{-1} (5·13–6·25 μ), but for maximum resolution of O—H, N—H and the various C—H stretching absorptions between 3,500 and 2,800 cm^{-1} (2·86–3·57 μ) a fluoride prism is necessary (*Figure 15*); a bromide prism is necessary below 650 cm^{-1}. It should be noted that almost all frequency ranges quoted for vibration modes have been established with sodium chloride optics.

Gratings allow better resolution than is obtainable with prisms; a grating which reflects light in the first order at a frequency of n cm^{-1} also reflects light in the second, third and fourth orders, according to the progression n, $2n$, $3n$, $4n$ . . . cm^{-1}. In practice, the unwanted orders are rejected by installation of a small prism, though filters may also be used. By linking the wavelength scan of the prism with that of the grating the spectrum may be taken, using a succession of orders from a single grating. Gratings offer a significant improvement in the signal : noise ratio and, in consequence, good resolution is obtainable at high recording speeds.

QUANTITATIVE ANALYSIS

COMPARISON of absorption intensities or determination of their individual values is the basis of quantitative analysis. The most reliable results are obtained by comparative methods since, despite much refinement of instrumental design and of operating procedures, reproducibility of absolute intensities of absorption on different instruments remains unsatisfactory.

Absorption Intensity

Beer's law and Lambert's law express, respectively, the relation of the intensity of absorption to changes in concentration and sample thickness. A general absorption law, which is a combination of Beer's and Lambert's laws, relates the absorption of incident monochromatic radiation by a substance dispersed in a non-absorbing medium to both concentration and sample thickness, and is expressed in equation (3) or, alternatively, written as equation (4)

$$I = I_0 10^{-kcd} \qquad . \quad . \quad . \quad (3)$$

$$kcd = \log_{10} I_0/I \qquad . \quad . \quad . \quad (4)$$

I and I_0 are the intensities of the transmitted and incident radiation, respectively, k is the extinction coefficient (absorptivity), c the concentration of the substance in g/l., and d the thickness of the sample in cm. Transmittance, T, is defined by equation (5), and percentage transmittance, $\%T = 100T$.

$$T = I/I_0 \qquad . \quad . \quad . \quad (5)$$

Arising from these equations are an abundance of terms used by various authors in the literature, summarized by BRODE[11]. Only a few of these terms will enter into this discussion. Absorbance, A, also termed extinction, E, or optical density, is given by $A = kcd$. If, as is becoming the custom, the concentration, c, is given in g mol./l., the symbol k is replaced by ϵ, and is now the molar extinction coefficient (molar absorptivity). A logarithmic reciprocal relation therefore exists between absorbance, A, and transmittance, T, given by $A = \log_{10}{}^{1}/T$. Matched solution and solvent reference cells effectively cancel out absorptions due to the solvent. However, in regions of strong absorption by the solvent, the transmitted light is very weak or zero and the signal from the detector is correspondingly small. The pen therefore responds only slowly, and regions of strong solvent absorption may not be used for spectral analysis of the sample (see pp. 28–32). This disadvantage is overcome by a constant energy slit control, instead of cam programming of slit width, to ensure a 'live' pen at low transmittance percentages. However, constant energy slit control requires very wide slit widths at low transmittance and may, in consequence, lead to marked alteration of band shapes and intensities because of the 'finite slit width' effect (p. 38).

Measurement of Intensity

In recent years instrument manufacturers have produced chart paper with a logarithmic ordinate scale which permits direct readings of absorbance instead of per cent transmittance. Calculations based on band heights give molar extinction coefficients which may be 'true' or 'apparent' depending on the conditions of measurement. For comparison of intensities of differently shaped bands, or in cases where finite slit errors (vide infra) are introduced, band areas should be used. Absorption band areas (integrated absorption intensities) may be evaluated by standard graphical procedure using a planimeter, by weighing paper profiles of the bands, or by counting squares. For the second procedure good quality paper of fairly uniform thickness is required, and, since the best results will be obtained only by semi-statistical methods, a rather tedious practical operation is incurred. Two mathematical approaches to the determination of integrated absorption intensities were developed[29,30] as alternatives to direct area measurement, and subsequently improved by RAMSAY[31] for solution spectra. Electrical band intensity integrators are becoming available for some instruments where the pen moves on a slide wire, which serves also as a potentiometer wire. The major advantage of determining integrated absorption intensities is their virtual independence of slit width, whereas measurements of absorbance or molecular extinction coefficient vary with slit width. ARNAUD[32] has reviewed in detail the subject of infra-red intensity measurements.

Quantitative Analysis of Mixtures (Determination of Purity)

The two principal procedures for quantitative analysis of mixtures, outlined earlier (p. 8), were based on the use of a percentage transmittance–concentration curve or on application of Beer's law. In each case greatest accuracy can be obtained by working with absorption bands of transmittance 25–65 per cent. These two procedures are applied generally to solution spectra. A similar approach has been developed[33] for quantitative analysis of solid mixtures examined spectroscopically by the alkali halide disc technique. Here an internal standard, potassium thiocyanate, is thoroughly mixed with powdered potassium bromide to obtain a large quantity of 1–2 per cent thiocyanate mixture. A series of spectra is then run of the sample at various concentrations in discs prepared from the thiocyanate–potassium bromide standard mixture, and the ratio of CNS^- ion absorption at 2,125 cm^{-1} (4·7 μ) to a chosen sample band absorption is plotted against percentage concentration of sample; a calibration curve is then obtained. Using the same thiocyanate–bromide mixture as standard disc material the concentration of the substance in any mixture may then be read off from the curve. A critical factor is the need for a constant grinding time in disc preparation. This procedure circumvents the need to measure the disc thickness and allows examination of substances which do not conform to Beer's law. Quantitative analysis using an internal standard in Nujol mulls has also been described[34].

Deviation from Beer's Law

Equations (3) and (4), and the expression $A = kcd$, are all mathematical representations of Beer's law; however, there are many substances which do not obey this relationship. Physical effects, particularly intermolecular association (e.g. hydrogen bonding) will cause deviations from Beer's law. Bands change shape and position and only the calibration curve method may be used for quantitative studies. Rigid control of operating conditions is vital; speed of scan, for example, should be slow and constant. Chemical effects such as dissociation and polymerization also lead to deviation from Beer's law.

A major cause of deviations is due to the 'finite slit width' effect, for no monochromator provides pure monochromatic light at the exit slit. Even with the highest quality optical surfaces, a prism or grating giving its maximum dispersion of the light, and narrow entrance and exit slits (all factors helping to 'purify' the light frequency), the light emerging from the exit slit will still cover a narrow frequency range. Well-constructed gratings, however, give light more nearly monochromatic than any prism can provide.

For a spectrophotometer frequency setting of n cm^{-1} the light emerging from the exit slit consists mainly of this frequency, but is contaminated by other frequencies ranging to several wave numbers on either side of this mean figure. Thus, at a frequency setting on $n - 2$ cm^{-1} some light of frequency n cm^{-1} will be present, and vice versa. When the spectral slit width is comparable to the half band width of an infra-red band, instead of being much smaller, the observed band shape deviates from true. The effect is to broaden the true absorption band, lower its height and, hence, reduce the molecular extinction coefficient (*Figure 16*).

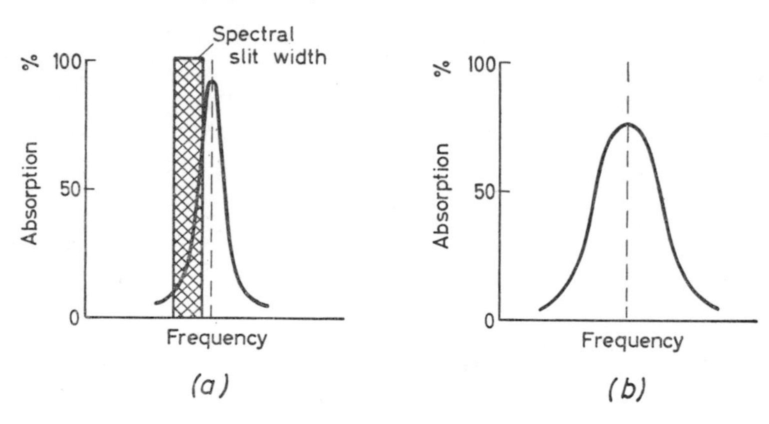

Figure 16. (a) True absorption band; (b) observed absorption band showing effect of finite spectral slit width

The individual operator can influence strongly the quality of a spectrum and the accuracy of absorption intensity data obtained therefrom, for selection of optimum settings of the instrument variables is the operator's responsibility; this is a critical factor in spectral analysis. Other factors, mentioned in earlier sections, which will materially influence the absorption intensity, include radiation losses due to scattering at interfaces and rough cell plate surfaces, stray light interference, ill-matched cells, solvent absorption, and non-standardization of sample preparation techniques. It is small wonder that, in general, absorption band intensities are denoted simply as weak, medium, strong or very strong!

At this point in the text it is pertinent to mention single-beam operation in spectroscopy. Though largely replaced by the more convenient double-beam instrument for routine infra-red spectral analysis, single-beam working eliminates or reduces many of the adverse factors in intensity measurements. Indeed, single-beam instruments still find extensive employment in both industrial and analytical laboratories and are irreplaceable for much fundamental research in spectroscopy.

Practical Applications of Intensity Measurements

In spite of the many difficulties emphasized above, a considerable amount of useful data on absorption band intensities has accumulated. In the unbranched alkanes the apparent molecular extinction coefficient of the methylene group is an additive function, increasing chain length causing increases in the value of ϵ by almost equal increments for each methylene group. This relation holds for CH_2 scissoring, wagging, and rocking vibration modes. Intensity studies of C—H stretching vibrations may be made in the 8,000–5,000 cm^{-1} region where the first overtones of these fundamental absorptions occur. High-sensitivity detectors may be used here in conjunction with high resolution monochromators. Absorption bands are separated further apart in the overtone region.

Secondary amides examined in dilute solution show an absorption band which may be resolved into a doublet probably corresponding to the *cis* and *trans* forms (V and VI). By measuring the ratio of the two absorption intensities the percentage of each stereoisomer present may be determined directly.

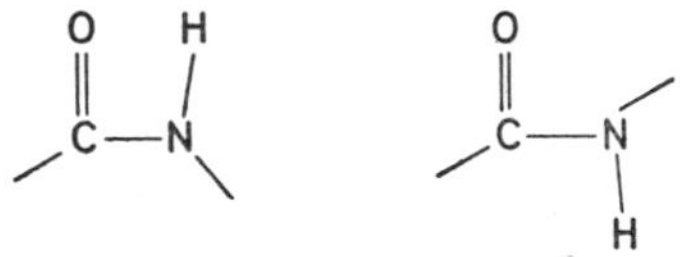

V

VI

Simple secondary amides have been shown, by this method, to exist largely in the *trans* form, with the *cis* form predominating in some sterically hindered compounds.

Jones and his co-workers have made a series of elegant studies on carbonyl integrated absorption intensities, which are summarized by Jones and Sandorfy[5], together with much other data on intensities. A combination of frequency and intensity measurement for any carbonyl absorption should lead to its complete identification.

HYDROGEN BONDING

INFRA-RED spectroscopy has found extensive use for studies of hydrogen bonds. A proton attached to a more electronegative atom X (e.g. Cl, O, N, P, Br, F, S) forms a partial bond with a neighbouring electronegative atom Y (e.g. Cl, O, N, P, S, Br, F) or *pi* bond (e.g. C=C, C=C, aromatic ring). Atom Y or the *pi* bond provides two electrons in an asymmetric orbital for the hydrogen bonding to develop. The strongest hydrogen bonds are formed where the atomic centres X, H and Y are collinear. In the infra-red spectrum hydrogen-bonded protons are characterized by shifts to lower frequencies (higher wavelengths) of the X—H stretching vibration mode, coupled with a marked increase in intensity of this absorption. Hydrogen bonding (association) involving —O—H groups causes the largest shifts, with lesser ones observed for —NH— groups. Only weak hydrogen bonds develop with S—H and P—H groups. Infra-red spectroscopy offers a simple method for distinguishing between intramolecular hydrogen bonding, intermolecular hydrogen bonding, and chelation (very strong intramolecular hydrogen bonding) such as that occurring in β-diketones. Individual factors, namely temperature, concentration and the sample diluent, can substantially alter the bonded O—H or N—H vibration absorption frequency positions. Therefore, standardization of operating conditions is a critical factor in comparative studies.

Generally, only when a compound is examined in the gas phase or in very dilute solution in a non-polar solvent will absolutely free (unassociated) O—H stretching absorption bands appear in the spectrum. Practical differentiation between the various types of hydrogen bonding is possible. Intramolecular hydrogen bonded, and chelated, O—H stretching absorption band frequencies are unaffected by dilution. In contradistinction, dilution of an intermolecularly hydrogen bonded hydroxylic compound in a non-polar solvent (other variables held constant) causes a reduction of the intermolecular bonding (*Figure 17*). This is manifested as a decrease in intensity of the absorption band for bonded OH stretching and a concomitant increase in the intensity of the free hydroxyl absorption. Since chelate compounds are readily detected by their very broad absorption, at substantially lower frequencies, a convenient practical method for hydrogen bond studies is available. Intramolecular hydrogen bonds are, by definition, confined to single molecules and the corresponding absorption band is sharp, except where chelation occurs. Conversely, intermolecular hydrogen bonds may be between two molecules (dimeric), when a fairly sharp infra-red absorption develops, or between several molecules (polymeric association), when a broad absorption band results.

Vibration absorptions other than A–H stretchings and bondings are affected by hydrogen bonding (e.g. C–O and C=O stretchings), and are referred to in the tables (Part II). Carbon disulphide or carbon tetrachloride, freed of hydroxylic impurities, are suitable solvents for hydrogen bond studies. A range of cells of varying path length (e.g. 0·05–2·5 mm) is required to facilitate examination of a wide concentration range (e.g. 1·0–0·02 molar).

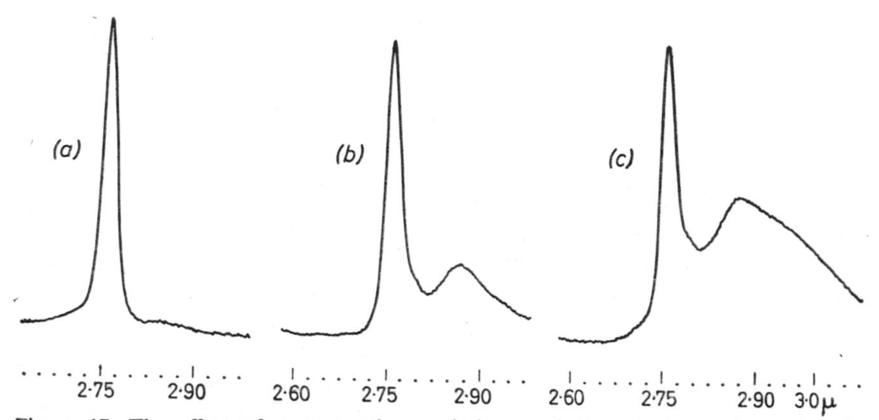

Figure 17. The effect of concentration variation on intermolecular hydrogen-bonded hydroxyl OH stretching absorption. Benzhydrol examined in carbon tetrachloride solution using a calcium fluoride prism in a Grubb-Parsons DB1 spectrophotometer: (a) 40 mg/20ml., 1 cm cell; (b) 40 mg/2ml., 0·1 cm cell; (c) 40mg/1ml., 0·05 cm cell. (*By courtesy of* R. L. Erskine)

For hydroxyl absorptions a lithium or calcium fluoride prism, or a grating, is preferred as the dispersion unit.

Several examples will demonstrate the applications of infra-red spectroscopy to stereochemical problems involving hydrogen bonding. BADGER[35], in a consideration of association involving the OH group, noted that the size of the frequency shift ($\Delta\nu$) from free hydroxyl absorption to associated hydroxyl absorption was a measure of the force constant of the OH · · · O— bond. Therefore $\Delta\nu$ should vary inversely with the length of the hydrogen bond. KUHN[36] examined in detail a large number of acyclic and cyclic diols in carbon tetrachloride solution at concentrations low enough to exclude the possible formation of *inter*molecular hydrogen bonds. The presence of an *intra*molecular hydrogen bond was detected by the appearance of a second OH band additional to that for free OH. The separation of these bonds, $\Delta\nu$, was taken as a measure of the length of the hydrogen bond. Only when the length of the hydrogen bond was less than 3·3Å, as calculated from a knowledge of bond lengths and bond angles, did Kuhn observe two OH bands. Thus, *cis-cyclo*hexane 1,2-diol showed a second OH band for which the downward displacement from the free hydroxyl O—H stretching frequency was 39 cm^{-1} and the H · · · O distance is 2·34 Å. *Trans-cyclo*hexane-1 : 2-diol can exist in diequatorial or diaxial forms with H · · · O distances of 2·34 Å and 3·3 Å, respectively. The infra-red spectrum showed $\Delta\nu =$ 32 cm^{-1} for the two OH stretching absorptions of this diol, from which it was concluded that the diol has the *trans* diequatorial stereochemistry. Kuhn also studied cyclohexane-1 : 3- and 1 : 4-diols and *cyclo*pentane diols. The expected downward shifts of C=O stretching absorption frequencies are observed whenever the oxygen atom is involved in association and the C=O bond perturbed. Substitution of the anthraquinone nucleus by hydroxyl results in a marked downward shift of the C=O stretching frequency when the hydroxyl is in positions 1-, 4-, 5- or 8-, since strong hydrogen bonds develop; hydroxyl substitution at the 2-, 3-, 6- and 7- positions does not affect this band[37].

INTERPRETATION OF A SPECTRUM

IN THE preceding sections discussion has centred on the problem of obtaining a reliable, well-resolved spectrum. Before interpretation of the spectrum is attempted the frequency (wavelength) accuracy must be known, either by pre-calibration of the instrument using ammonia gas, water vapour, indene or polystyrene as references, or by calibration of individual spectra, usually by means of a polystyrene film. The latter procedure requires only a minute or so on most instruments and should be a standard operation with low-cost instruments, or whenever single chart papers are used, since the positioning of the paper on the recording drum constitutes a source of error. Polystyrene calibration (e.g. *Figures 11–14*) covers effectively the range 4,000–650 cm^{-1} (2·5–15·4 μ). Positions of the principal bands, in cm^{-1} and μ units, are presented in *Table 7*; and the complete spectrum is included here (*Figure 18*).

An authoritative book dealing with the accurate calibration of prism and grating infra-red spectrometers is now available[38].

Table 7. Principal Absorption Bands of Polystyrene Film

Frequency (cm^{-1})	Wavelength (μ)	Frequency (cm^{-1})	Wavelength (μ)
3,062	3·266	1,602	6·243
3,027	3·303	1,494	6·692
2,924	3·420	1,154	8·662
2,851	3·508	1,028	9·724
1,944	5·144	907	11·03
1,802	5·549	700	14·29

No rigid rules exist for interpretation of a spectrum, but certain general observations are helpful. Appearance of an absorption band where predicted on the basis of prior knowledge of the compound should not, in itself, be regarded as conclusive evidence for the existence of a group. Interference by other absorptions must be eliminated from the possibilities, and positive evidence sought by examination of other regions of the spectrum. Conversely, absence of a strong group absorption is usually indicative of the absence of that group in the molecule, provided no effects are operating (e.g. hydrogen bonding) which could shift the absorption band to other regions. Thus, absence of strong absorption in the region 1,850–1,640 cm^{-1} (5·40–6·50 μ) excludes carbonyl groups from the molecular structure. Preliminary examination of the spectrum should definitely concentrate on the regions above 1,350 cm^{-1} (below 7·40 μ) and 900–650 cm^{-1} (11·1–15·4 μ). The intervening zone, 1,350–900 cm^{-1} (7·4–11·1 μ), the so-called 'fingerprint' region, frequently comprises a large number of bands whose origin is not easily determined. Nevertheless it is a useful source of information, particularly when studied with reference to bands in the lower and higher ranges. Furthermore, total interpretation of a spectrum is seldom required, much valuable structural evidence being attainable from relatively few bands. Absorption bands

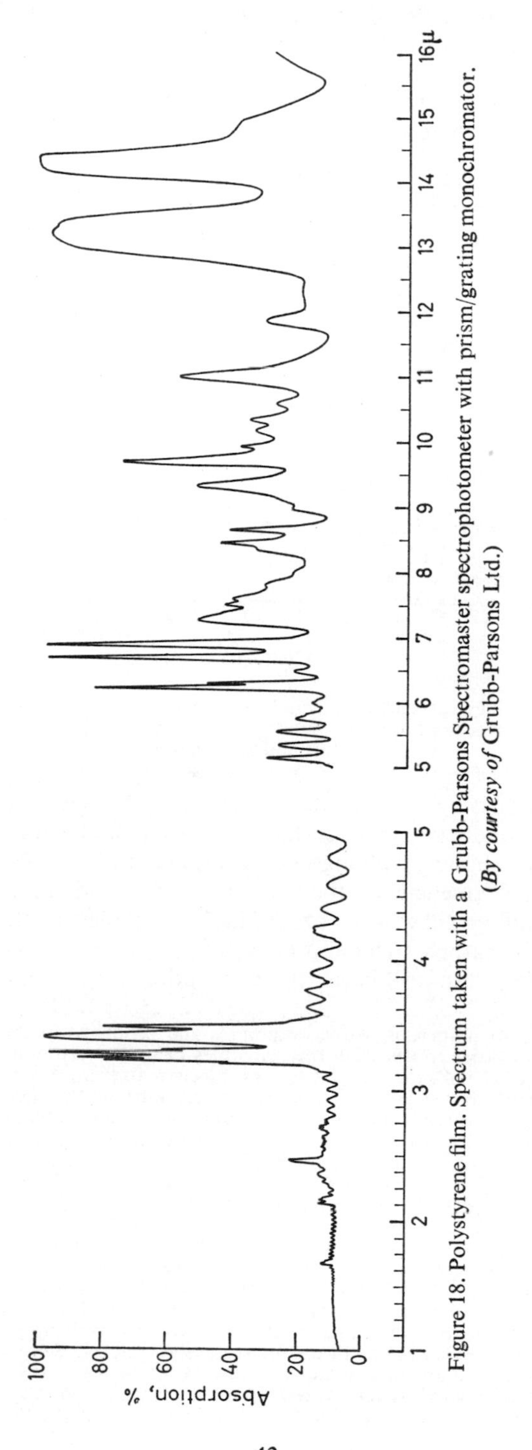

Figure 18. Polystyrene film. Spectrum taken with a Grubb-Parsons Spectromaster spectrophotometer with prism/grating monochromator.
(*By courtesy of* Grubb-Parsons Ltd.)

in regions of strong sample diluent absorption are meaningless for deductions concerning molecular structure of the sample*.

Assignment of bands to specific groups is facilitated by use of isotopes or chemical changes. Deuterium exchange is particularly helpful for assignment to A–H vibrations, where the hydrogen is exchangeable. Moreover the band

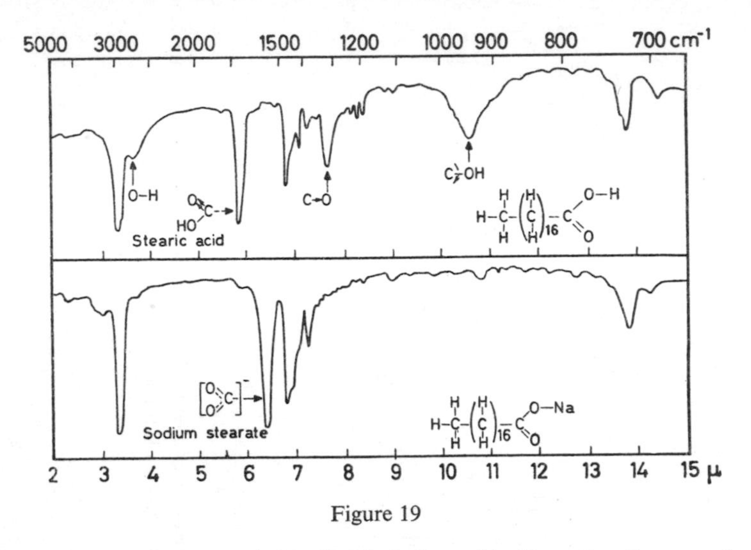

Figure 19

Figures 19–21. Samples suspended in Nujol. Sodium chloride optics. Spectra taken with an early Beckman spectrophotometer model IR-2T. (*By courtesy of* J. M. Vandenvelt, R. B. Scott and Beckman Co.)

frequency shifts arising from deuterium isotope substitution are calculable on the basis of Hooke's law (p. 2). Thus, it can be shown that the absorption frequencies for a bond involving deuterium are, to a rough approximation, $1/\sqrt{2}$ times the frequencies of the corresponding bonds involving hydrogen. Hydrogen bonding may also be studied by replacement of proton by deuterium. This mathematical approach fails when the A–H (or A–D) vibration is not solely responsible for the absorption, as in combination bands.

* The provision of numerous worked examples and problems in the text at this point may be argued, in order to illustrate the principles behind interpretation. However, it is the author's opinion that for the beginner the best place to study infra-red spectral interpretation is undoubtedly in the laboratory under expert tuition. Problem–answer examples written by an expert conscious of the relative values of assignments, although potentially valuable, are fraught with pitfalls for the unsuspecting student. In practice, it is only seldom that one can derive with certainty the complete structure of a compound by infra-red spectroscopy alone. Yet, how frequently are problems styled as 'What is the structure of compound X, whose infra-red spectrum is shown below?' In an effort to squeeze the maximum of information from a spectrum it is tempting to make assignments and structural correlations for bands of a highly speculative nature, which is to be strongly discouraged. Every worthwhile University or College undergraduate course in organic analysis should provide the student with an infra-red spectrum of the unknown compound. The spectrum can be discussed with a demonstrator and examined *in conjunction with* chemical and other spectroscopic evidence. For the student who is unable to participate in such a course attendance is recommended at one of the first-rate Spectroscopy Seminars now being held at regular intervals in both Europe and the United States. Once the initial problems of interpretation are conquered students will undoubtedly benefit from working over numerous spectra, and a most timely contribution to this end appeared in 1962[9].

The $^{12}C{\equiv}N$ stretching vibration absorption at 2,213 cm^{-1} shifts to 2,157 cm^{-1} for $^{13}C{\equiv}N$ which is within 1 cm^{-1} of the theoretical shift.

Conversion of an acid to its salt (*Figure 19*), an ester, or its primary amide (*Figure 20*) permits assignment of several bands to the carboxyl group. Similarly, an amino acid is readily converted to its hydrochloride or

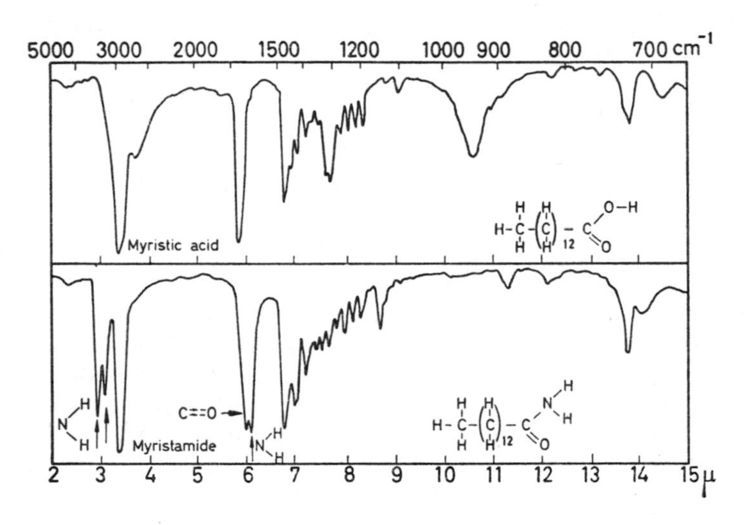

Figure 20

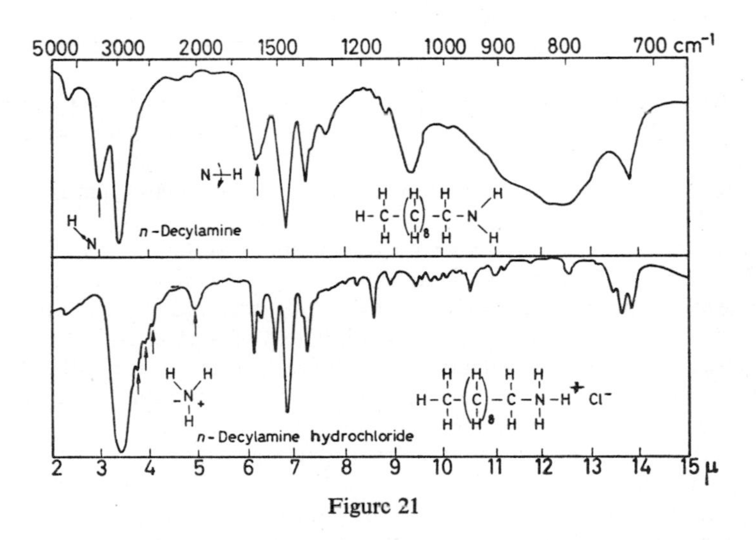

Figurc 21

metal salt, corresponding to the change $H_3\overset{+}{N}—C—CO_2^- \rightarrow H_3\overset{+}{N}—C—CO_2H$ or $H_2N—C—CO_2^-$, when several major infra-red absorption bands disappear, or shift, for each change; the spectral changes stemming from the conversion of an amine to its hydrochloride are illustrated in *Figure 21*. Bands for ethylenic linkages vanish when a compound is hydrogenated. The above examples outline the value of simple chemical changes to the interpretation of a spectrum.

A useful technique for spectral examination of crystalline solids and fibres involves the use of polarized infra-red radiation, obtained by reflection at selenium mirrors or by transmission through inclined plates of selenium or silver chloride. Plane polarized radiation is passed through the mounted crystals or fibres in two directions at right angles, when the intensity changes of characteristic bands may be correlated with the overall molecular structure. Absorption will be strongest when the vibrating dipole is aligned in the same plane as the polarized radiation and is zero when the dipole is at right angles to this plane. In practice the dipoles cannot be oriented to lie in these two desirable positions for maximum and zero absorption, but lie at a skew-angle to the plane of polarization and are examined in two mutually perpendicular planes. Proteins and polypeptide chains, plastics and rubber have all been usefully studied by this technique.

Substantial aids to identification of unknown compounds are provided by several reference collections of infra-red spectra, which may be general in their scope or limited to specific fields. Thus, an atlas of 760 steroid infra-red spectra is available[39, 40], and 354 spectra, mainly of compounds studied in connection with the structural elucidation of penicillin, have been assembled[8].

Under the auspices of the American Petroleum Institute (Research Project 44) there have been gathered together, on loose sheets suitable for filing, the infra-red spectra of many hundreds of hydrocarbons. Jones and Sandorfy list nine papers where the infra-red spectra of some highly specialized groups of compounds are catalogued[41].

In book form there are two smaller general compilations dealing respectively with organic compounds[6] and a selection of both organic and inorganic compounds[42].

On the recommendation of its Infra-red Absorption Data Joint Committee, the Chemical Society, London, set up the D.M.S. (Documentation of Molecular Spectroscopy) Index[43]. Approved Spectra are recorded on punch cards together with the frequencies of the eight strongest bands between 5,000 and 665 cm^{-1} (2–15 μ).

Similarly, the American Society for Testing Materials (A.S.T.M.) in association with Wyandotte Chemicals Corp., has catalogued on punched cards serial numbers which refer to a source of infra-red spectral absorption data for many thousands of compounds[44]. The cards can be sorted on I.B.M. machines to correlate spectral data with chemical structure; spectral data from the other major collections is included on the A.S.T.M. cards. Also using the A.S.T.M. system for coding are two punched card collections organized by the U.S. National Bureau of Standards in co-operation with the National Research Council[45]. One card carries the spectrum of the compound and a survey of the pertinent literature. In the parallel set (bibliography cards) the corresponding card carries an abstract of the paper appearing in the literature.

The immense Sadtler collection of infra-red spectra, exceeding 23,000 multiply-indexed organic compounds, has been placed on I.B.M. cards and magnetic tapes[46]. A further 1,850 spectra are added annually to this collection. Every spectrum carries an identification number and is classified by

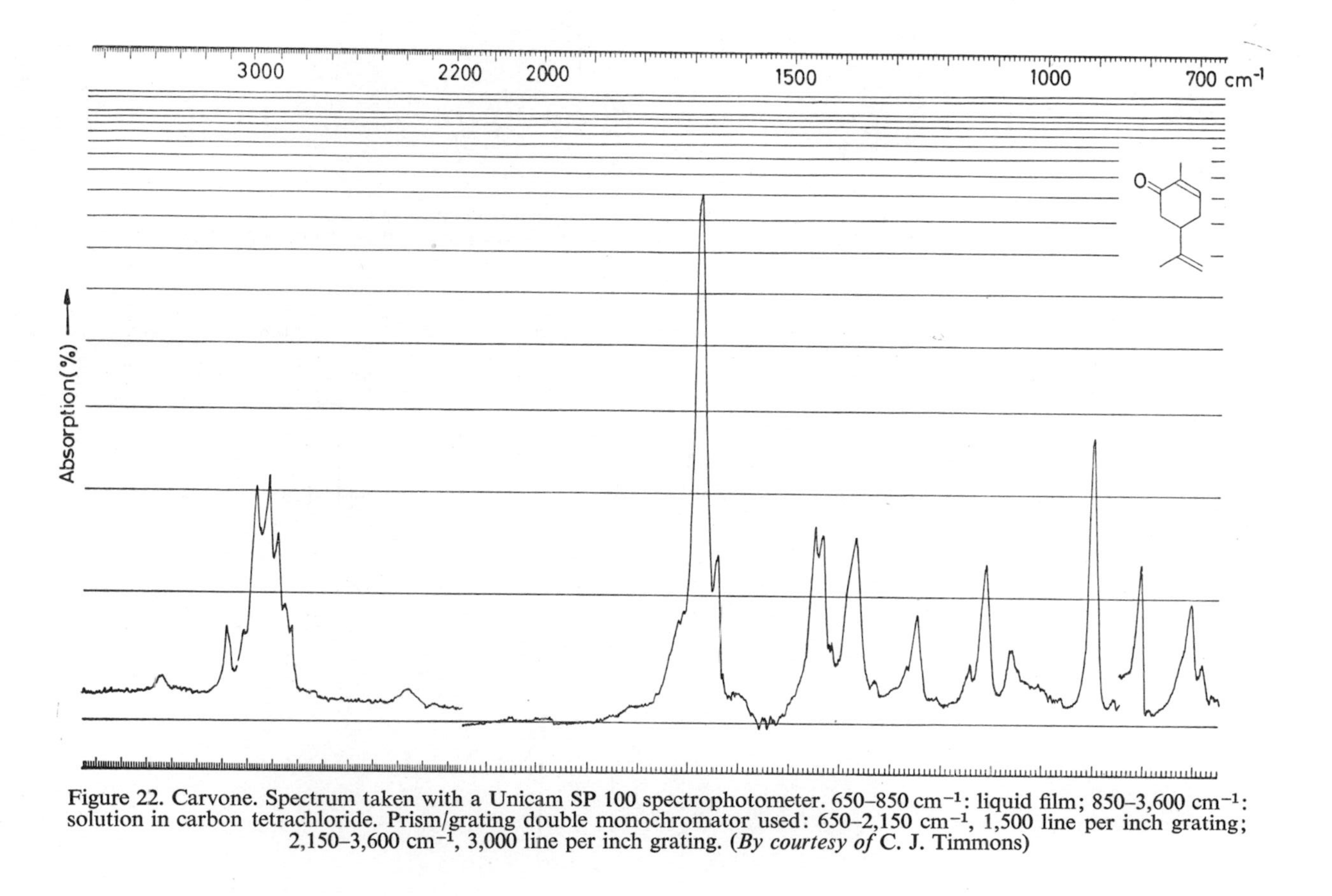

Figure 22. Carvone. Spectrum taken with a Unicam SP 100 spectrophotometer. 650–850 cm^{-1}: liquid film; 850–3,600 cm^{-1}: solution in carbon tetrachloride. Prism/grating double monochromator used: 650–2,150 cm^{-1}, 1,500 line per inch grating; 2,150–3,600 cm^{-1}, 3,000 line per inch grating. (*By courtesy of* C. J. Timmons)

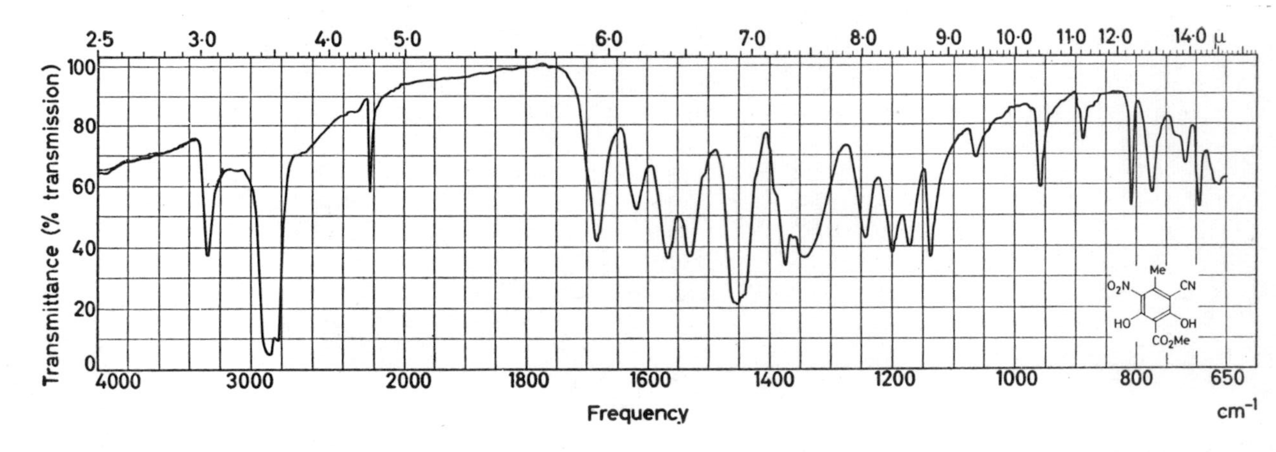

Figure 23. Methyl 3-cyano-2,6-dihydroxy-4-methyl-5-nitrobenzoate. Spectrum taken with a Perkin-Elmer 21 spectrophotometer using a sodium chloride prism and a Nujol mull. (*By courtesy of* Professor D. H. R. Barton)

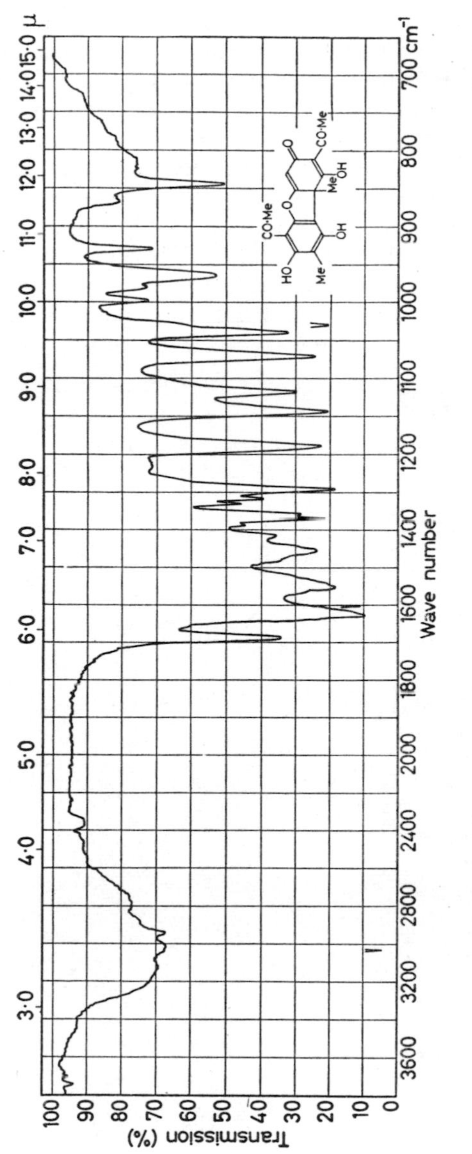

Figure 24. Usnic acid. Spectrum taken with a Nihon Bunko Co. Koken DS-301 spectrophotometer with sodium chloride prism, using a solution in chloroform. (*By courtesy of* Professor S. Shibata)

compound type, melting or boiling point, molecular formula, and also by the strongest absorption bands for each 1 μ between 2 and 14 μ, each band being coded to the nearest 0·1 μ division. Data for an unknown compound can now be processed through a computer and rapidly compared with all the available data. Another card index of spectra is also being compiled under the guidance of JONES[47]. An index of infra-red spectra up to 1957 has been published[48]. In 1962, an important new addition to the infra-red spectroscopy literature was made by Professor NAKANISHI[9]. In this book attention is drawn to a Japanese collection of card-indexed infra-red spectral data[49] and other useful references.

Part I is concluded with three complex spectra of compounds containing abundant and various functional groups. While a number of bands can be assigned quickly in each case there remain multiple absorptions for which structural correlations can be only tentative. These examples emphasize that it is often not possible (nor even desirable!) to attempt the assignment of all bonds in a spectrum. Yet valuable information is obtainable from each curve. Students are invited to discuss these spectra with tutors.

REFERENCES—PART I.

[1] DUNCAN, A. B. F., *Chemical Applications of Spectroscopy*; Interscience: New York, 1956.

[2] WILSON, E. B., DECIUS, J. C. and CROSS, P. C., *Molecular Vibrations—the Theory of Infra-red and Raman Spectra*; McGraw-Hill: London, 1955.

[3] HERZBERG, G., *Infra-red and Raman Spectra of Polyatomic Molecules*; van Nostrand: New York, 1945.

[4] CONN, T. I. and AVERY, D. G., *Infra-red Methods, Principles and Applications*; Academic Press: New York, 1960.

[5] JONES, R. N. and SANDORFY, C., in *Technique of Organic Chemistry*, Vol. IX, *Chemical Applications of Spectroscopy*, Ed. A. WEISSBERGER; Interscience: New York, 1956.

[6] BARNES, R. B., GORE, R. C., LIDDEL, U. and VAN ZANDT WILLIAMS, *Infra-red Spectroscopy*; Reinhold: New York, 1944.

[7] BELLAMY, L. J., *The Infra-red Spectra of Complex Molecules* (2nd edn); Methuen: London, 1958.

[8] RANDALL, H. M., FOWLER, R. G., FUSON, N. and DANGL, J. R., *Infra-red Determination of Organic Structures*; van Nostrand: New York, 1949.

[9] NAKANISHI, K., *Infra-red Absorption Spectroscopy*; Holden-Day, Inc.: San Francisco, 1962.

[10] JONES, R. N. and SANDORFY, C., in *Technique of Organic Chemistry*, Vol. IX, *Chemical Applications of Spectroscopy*, Ed. A. WEISSBERGER; Interscience: New York, 1956, p. 332.

[11] BRODE, W. R., *J. opt. Soc. Amer.*, 1949, **39**, 1022.

[12] RANDALL, H. M., FOWLER, R. G., FUSON, N. and DANGL, J. R. *Infra-red Determination of Organic Structures*; van Nostrand: New York, 1949, p. 11.

[13] RAMIREZ, F. and PAUL, A. P., *J. Amer. chem. Soc.*, 1955, **77**, 1035.

[14] SHEINKER, Yu. N. and RESNIKOW, V. M., *Dokl. Akad. Nauk S.S.S.R.*, 1955, **102**, 109.

[15] GIBSON, J. A., KYNASTON, W. and LINDSEY, A. S., *J. chem. Soc.*, 1955, 4340.

[16] BELLAMY, L. J. and WILLIAMS, R. L., *Spectrochim. acta*, 1957, **9**, 341.

[17] HOUGHTON, R. P., *Ph.D. Thesis*, London, 1958.

[18] GAUTHIER, G., *Compt. rend.*, 1951, **233**, 617.

[19] JONES, R. N. and SANDORFY, C., in *Technique of Organic Chemistry*, Vol. IX, *Chemical Applications of Spectroscopy*, Ed. A. WEISSBERGER; Interscience: New York, 1956, p. 327.

[20] *Chem. & Engng News*, 1959, **37**, 44.

[21] BARKER, S. A., BOURNE, F. J., NEELY, W. B. and WHIFFEN, D. H., *Chem. & Ind.*, 1954, 1418.

[22] MUNDAY, C. W., *Nature, Lond.*, 1949, **163**, 443.

[23] JONES, R. N. and SANDORFY, C., in *Technique of Organic Chemistry*, Vol. IX, *Chemical Applications of Spectroscopy*, Ed. A. WEISSBERGER; Interscience: New York, 1956, p. 298.

[24] FARMER, V. C., *Spectrochim. acta*, 1957, **8**, 374; *Chem. & Ind.*, 1955, 586.

[25] GORE, R. C., BARNES, R. B. and PETERSEN, E., *Anal. Chem.*, 1949, **21**, 382.

[26] GOULDEN, J. D. S., *Spectrochim. acta*, 1959, **9**, 657.

[27] ANGELL, C. L., KREUGER, P. J., LAUZON, R., LEITCH, L. C., NOACK, K., SMITH, R. J. D. and JONES, R. N., *Spectrochim. acta*, 1959, **11**, 926.

[28] TARBELL, D. S., CARMAN, R. M., CHAPMAN, D. D., CREMER, S. E., CROSS, A. D., HUFFMAN, K. R., KUNSTMANN, M., MCCORKINDALE, N. J., MCNALLY, J. G. (JNR.), ROSOWSKY, A., VARINO, F. H. L. and WEST, R. L., *J. Amer. chem. Soc.*, 1961, **83**, 3096.

[29] WILSON, E. B. and WELLS, A. J., *J. chem. Phys.*, 1946, **14**, 578.

[30] BOURGIN, D. G., *Phys. Rev.*, 1927, **29**, 794.
[31] RAMSAY, D. A., *J. Amer. chem. Soc.*, 1952, **74**, 72.
[32] ARNAUD, P., *Bull. Soc. Chim. France*, 1961, 1037.
[33] WIBERLEY, S. E., SPRAGUE, J. W. and CAMPBELL, J. E., *Anal. Chem.*, 1957, **29**, 210.
[34] BARNES, R. B., GORE, R. C., WILLIAMS, E. F., LINSLEY, S. G. and PETERSON, E. M., *Ind. Engng Chem. Anal.*, 1947, **19**, 620.
[35] BADGER, R. M., *J. chem. Phys.*, 1940, **8**, 288.
[36] KUHN, L. P., *J. Amer. chem. Soc.*, 1952, **74**, 2492.
[37] FLETT, M. St. C., *J. chem. Soc.*, 1948, 1441.
[38] *Tables of Wavenumbers for the Calibration of Infra-red Spectrometers;* I.U.P.A.C. Commission on Molecular Structure and Spectroscopy; Butterworths: London, 1961.
[39] DOBRINER, K., KATZENELLENBOGEN, E, R. and JONES, R. N. *Infra-red Absorption Spectra of Steroids. An Atlas*, Vol. I.; Interscience: New York, 1953.
[40] ROBERTS, G., GALLAGHER, B. S. and JONES, R. N., *Infra-red Absorption Spectra of Steroids. An Atlas*, Vol. II; Interscience: New York, 1958.
[41] JONES, R. N. and SANDORFY, C., in *Technique of Organic Chemistry*, Vol. IX, *Chemical Applications of Spectroscopy*, Ed. A. WEISSBERGER; Interscience: New York, 1956.
[42] EUCKEN, A. and HELLWEGE, K. H., *Landolt-Bornstein Zahlenwerte und Funktionen aus Physik, Chemie, Astronomie, Geophysik und Technik*, Sechste Auflage, 1. Band, Atom und Molekular Physik, 2 Teil, Molekeln I. 328–478; 3 Teil, Molekeln II, 557–656; Springer Verlag: Berlin, 1951.
[43] *Documentation of Molecular Spectroscopy Index;* Butterworths: London, and Verlag Chemie: Weinheim.
[44] *Wyandotte–ASTM Cards;* American Society for testing Materials: Philadelphia.
[45] Infra-red Spectral and Bibliographic Punch Cards; National Research Council, Committee on Infra-red Absorption Spectra, National Bureau of Standards: Washington.
[46] *Sadtler Catalog;* Sadtler Research Laboratories: Philadelphia.
[47] Coblenz Society Cards; Dr. H. Keseler, c/o Perkin Elmer Co., Norwalk, Connecticut, U.S.A.
[48] *An Index of Published Infra-red Spectra;* Ministry of Aviation, H.M.S.O.: London, 1960.
[49] IRDC Cards: Infra-red Data Committee of Japan (S. Mizushima) Nankodo Co., Haruki-cho, Tokyo, Japan.

PART II

INTRODUCTION

THE second half of the book is composed largely of correlation charts and tables of group absorption frequencies. These provide the basic information necessary for a reasonable interpretation of the infra-red spectrum of an organic compound. Reciprocal tables and a general index are included. *It must be emphasized most strongly that incorrect use of these charts and tables will inevitably lead to erroneous conclusions.* Furthermore, since the tables give ranges of absorption frequencies for each structural unit, detailed spectral information for a compound or a related model should be sought in the literature. BELLAMY[1] and JONES and SANDORFY[2] have made comprehensive collections of references which are invaluable aids in this respect. Assignment of absorption frequencies to specific vibrations should be made only after a careful consideration of the variables involved, as described in Part I.

Most of the absorptions useful to the organic chemist have been included, as well as some involving phosphorus, boron, silicon, and inorganic ions which may be of wider interest. In general, all the absorptions quoted fall within the range 4,000–600 cm^{-1} (2·5–16·7 μ) since, though many instruments are capable of operating satisfactorily over greater ranges, relatively few structural assignments have been firmly established outside these limits. Current research in theoretical and practical spectroscopy suggests that an extension to a practical range of 10,000–300 cm^{-1} (1–33 μ), with many new valuable correlations, should be possible within a decade.

Following this introduction is a list of abbreviations employed extensively throughout the tables and correlation charts. Six correlation charts cover the range 3,700–600 cm^{-1} (2·70–16·67 μ); they are, essentially, a variation of the several correlation charts which have appeared in the literature since 1943, and their linearity in frequency, rather than wavelength, should be noted. Some overlap of the frequency range covered by individual charts has been introduced, allowing certain major vibration modes to appear on a single chart. *These charts should be used only in conjunction with the greater detail provided in the tables*, and the page numbers of relevant tables are given in the right-hand column. Attention is drawn to the method of representing several bands with close or overlapping ranges all due to one structural unit—successive band frequency ranges, with their intensities, are demarcated first above and then below the line of frequency range.

A logical sequence of bond types is followed in the tables; C–C and C–H, C–O, O–H, C–N, N–H, N–O, C–halogen, C–S, C–P, C–Si, and inorganic ions. For each bond type single-bond absorptions are tabulated prior to multiple-bond absorptions. Each table is divided into five columns containing, from left to right: the absorbing group; absorption frequency range, in units of cm^{-1}; absorption wavelength range, in units of μ; intensity of absorption; and, in the last column, various information on the absorption.

Horizontal subdivision of tables corresponds to a change in group or type of vibration.

The absorbing group is described by formulae or written group names according to the dictates of brevity. All information in the remaining columns then refers to this absorbing group until another name or formula appears in the first column. Thus, the vinyl group has no fewer than seven absorptions due to C–H stretching or deformation vibrations while a *cis* disubstituted olefinic link has three such absorptions (p. 64).

Frequencies are quoted to the fourth figure in the literature but wavelengths usually only to the third (below 10 μ); this anomaly is not repeated here. It is beyond the working capacity of several types of instruments, and of many operators, to provide accurate and reproducible frequency values where the fourth figure is meaningful. Accordingly the minimum frequency difference employed in these tables is 5 cm^{-1} below 2,000 cm^{-1} and, with a few exceptions, 10 cm^{-1} above that figure; wavelengths are quoted to the second decimal place. Even these margins can be optimistic for the simplified, small spectrophotometers, unless constant checks and corrections for errors are made. Discussions with several spectroscopists revealed a variance of opinion on the breadth of the frequency range which should be quoted for a given grouping. The fact that organic chemists, as a whole, work with a greater range of compounds than do research spectroscopists within their narrower fields leads to the quotation in these tables of frequency and wavelength ranges which are tolerably broad. A large percentage of absorption frequencies will fall in a narrower region near the centre of the quoted range, but molecular structure variations, with their associated steric or electrical effects upon the vibrating group, can cause marked absorption frequency shifts. External environment changes (e.g. solvent or phase) may similarly affect the vibration. Consequently, absorption frequency shifts may be of such magnitude as to remove the absorption well outside the quoted range. Some absorption frequencies are expressed as a single frequency preceded by the abbreviation *ca.* This indicates that insufficient examples have been studied to validate quotation of a frequency range, or that practical difficulties (e.g. weak intensities, interfering absorptions) hinder collection of data to establish a range. The abbreviation l.v. in the last column infers the limited value of this absorption frequency in structural analysis. In a very few cases a single frequency is given with no qualifying abbreviations, indicating that all such absorptions occur within ± 2 cm^{-1} of the quoted value. These absorptions, especially when of strong intensity, are invaluable. All the above remarks concerning frequency apply likewise to wavelength ranges given in the centre column of the tables.

In spite of intensive research, systematic intensity measurements are still limited to selected groups and vibration types. Practical difficulties of intensity reproducibility in passing from one instrument to the next, and even the uncertainty of obtaining consistent intensity values on the same instrument, allow the more recent (post-1950) literature intensities an accuracy of perhaps ± 20 per cent. Earlier quotations of intensities should generally be used only with reservation. Again, some divergence of view exists among spectroscopists. Intensities are therefore classified throughout the tables as weak (w), medium (m) or strong (s). NAKANISHI[4] has recently compiled tables of characteristic infra-red absorption frequencies with ranges of apparent molecular extinction coefficients where possible. Generally, this intensity data is available only for vibrations of X—H and certain C=Y bonds.

It should be noted that hetero atoms, some unsaturated links, and aromatic rings generally give rise to strong absorption bands. A few bands have very strong intensities (v.s.) considerably exceeding the intensity of an absorption normally described as strong. When the intensity of an absorption varies widely the abbreviation v is indicated.

The right-hand column is intended to provide additional information about the absorption, its origin, and particularly to specify any restrictions in its use for diagnostic purposes. Inconsistent band (i.b.) refers to group absorptions which cannot always be detected for every molecule containing the relevant structural feature. Limited value absorptions (l.v.), indicative of an insufficiency of reliable information, should be used only with the greatest caution and reservation.

ABBREVIATIONS

adj.	adjacent
approx.	approximately
asym.	asymmetrical
conj.	conjugated
def.	deformation
dil.	dilute
enh.	enhanced
i.b.	inconsistent band
int.	intensity
i.p.	in-plane
liq. ph.	liquid phase
l.v.	limited value assignment
m.	medium intensity
non-conj.	non-conjugated
o.o.p.	out-of-plane
'5' ring	5-membered ring, etc.
s.	strong intensity
sat.	saturated
sec.	secondary
soln.	solution
so. ph.	solid phase
spec.	spectrum
str.	stretching
sym.	symmetrical
tert.	tertiary
unsat.	unsaturated
v.	variable intensity
vap. ph.	vapour phase
vib.	vibration
v.s.	very strong intensity
w.	weak intensity

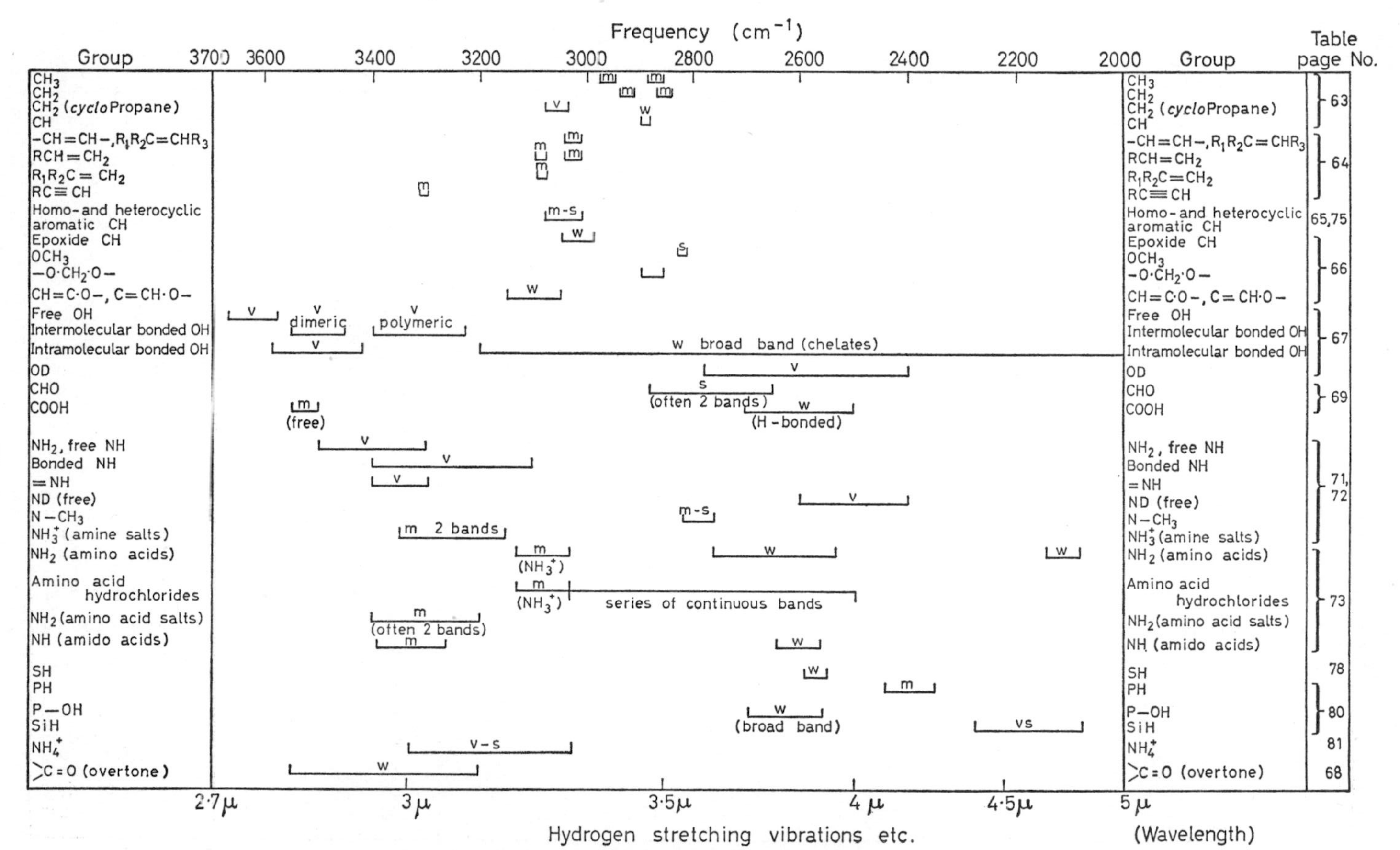

Correlation Chart I

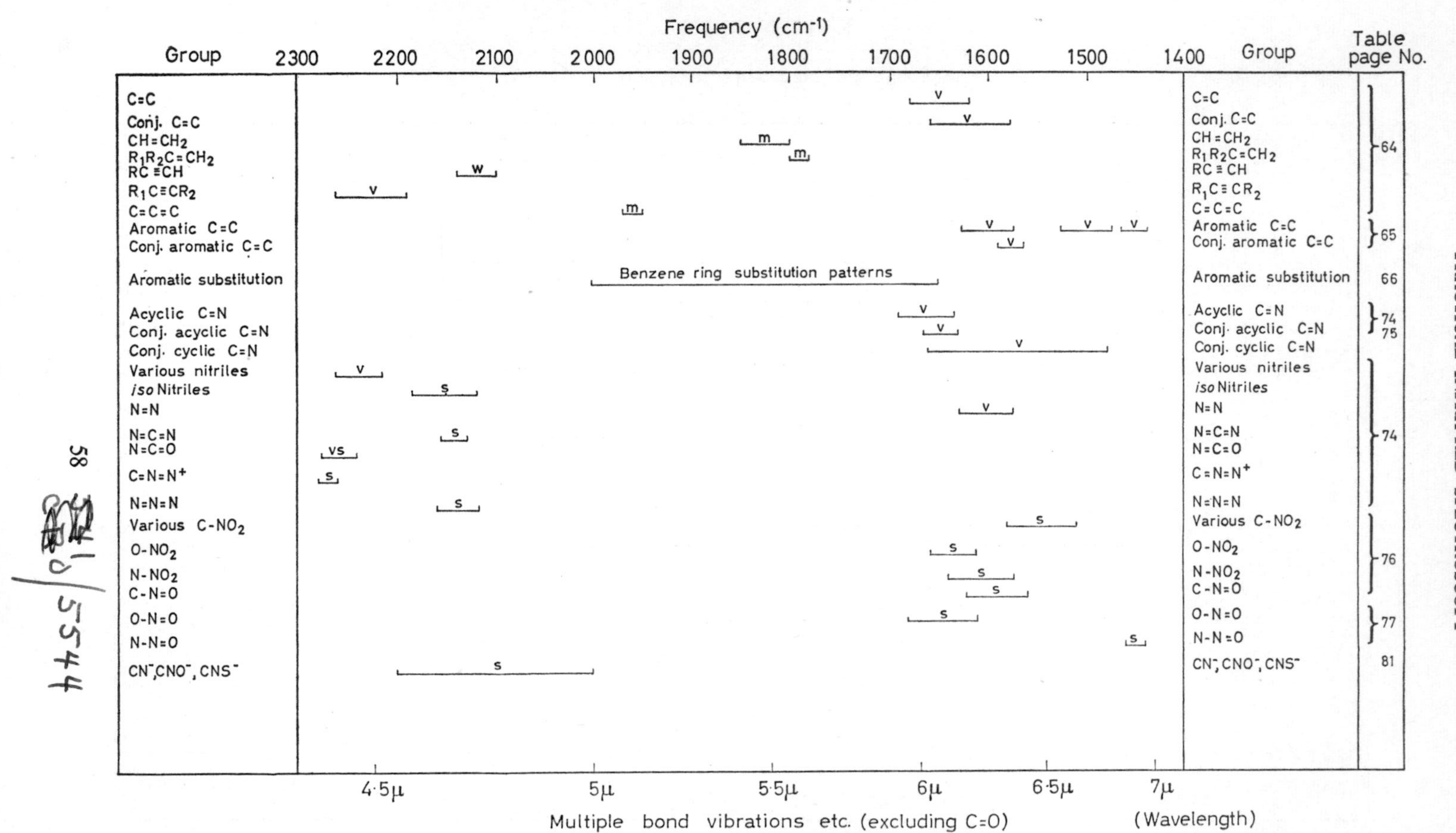

Correlation Chart II

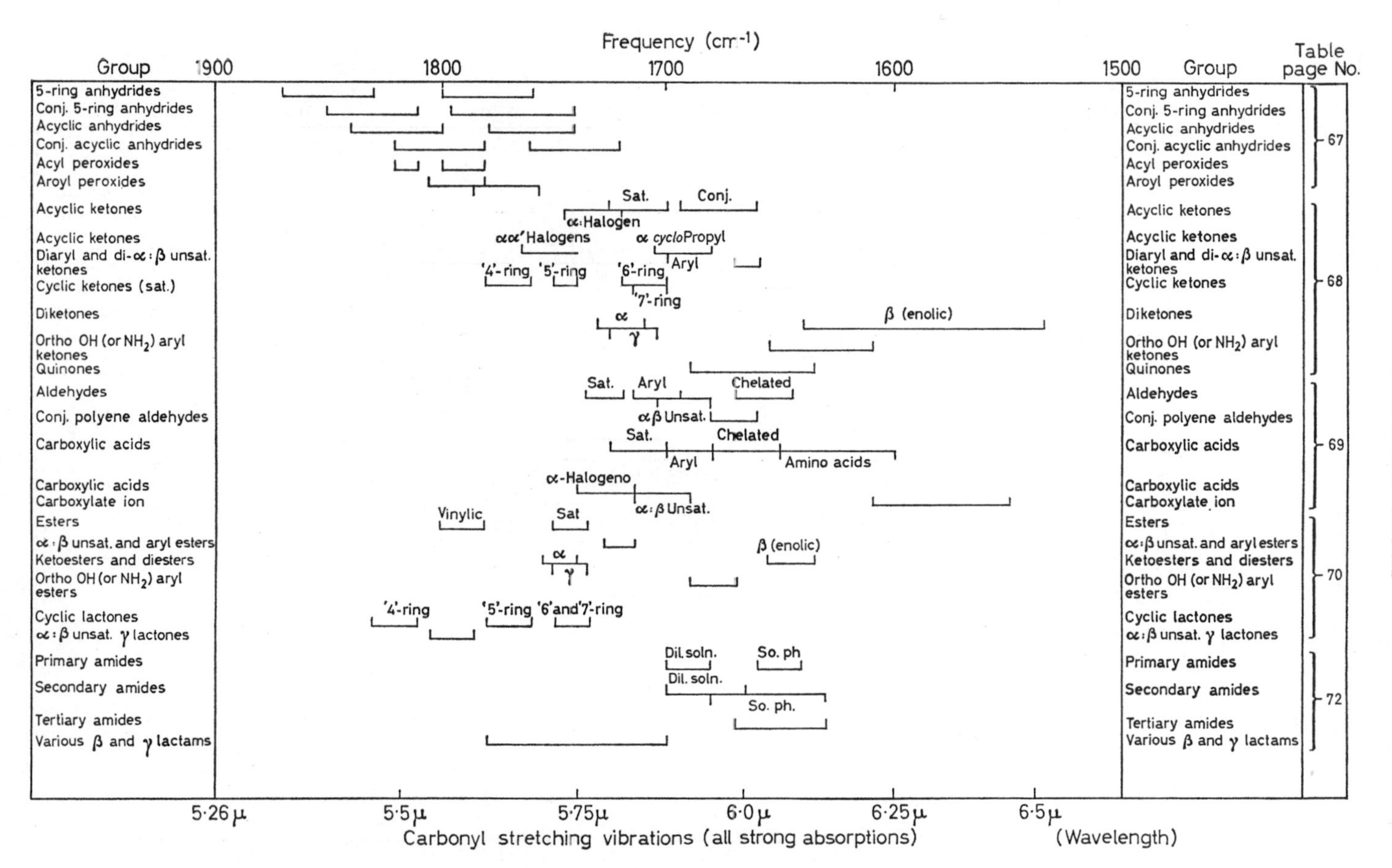

Correlation Chart III

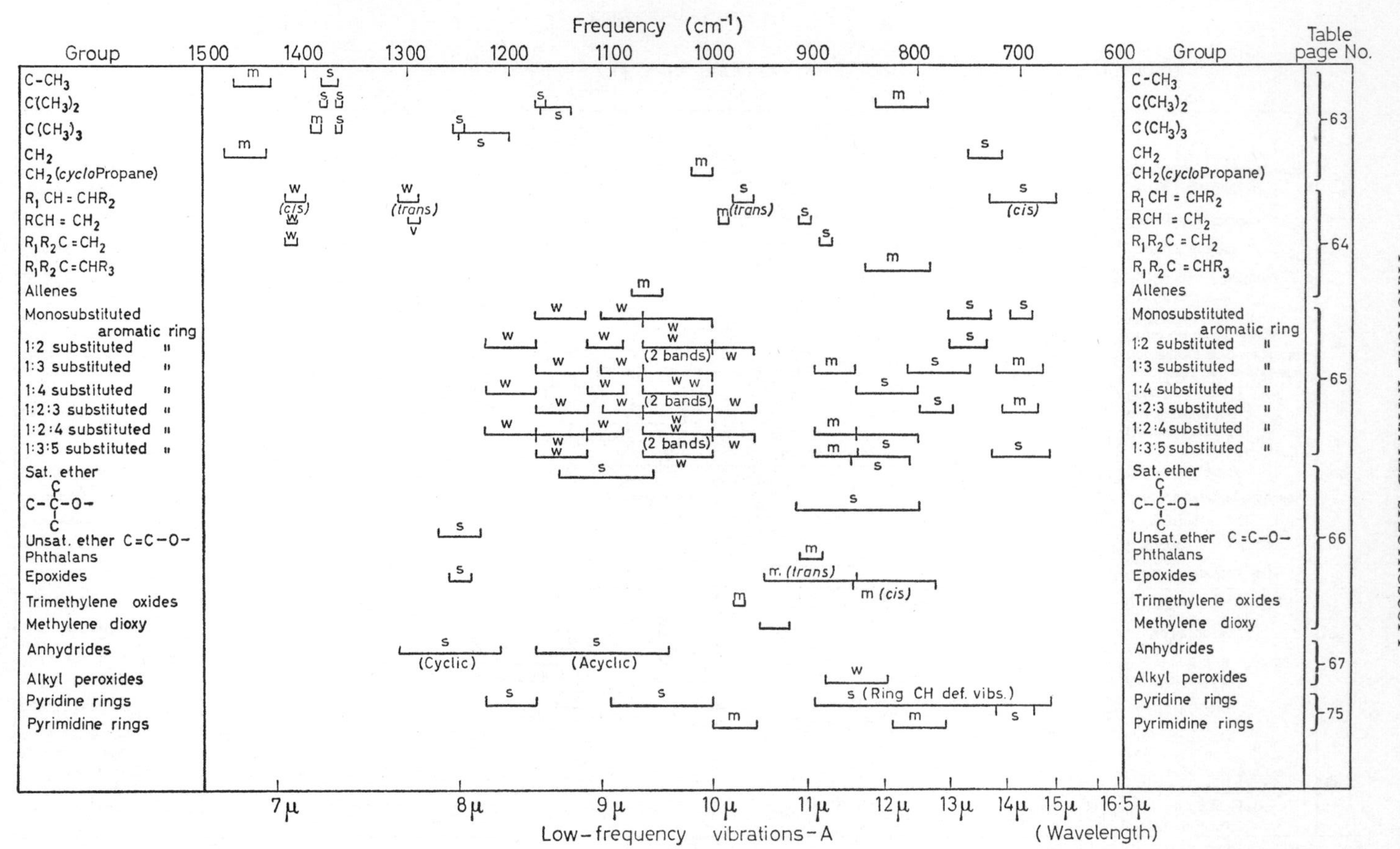

Correlation Chart IV

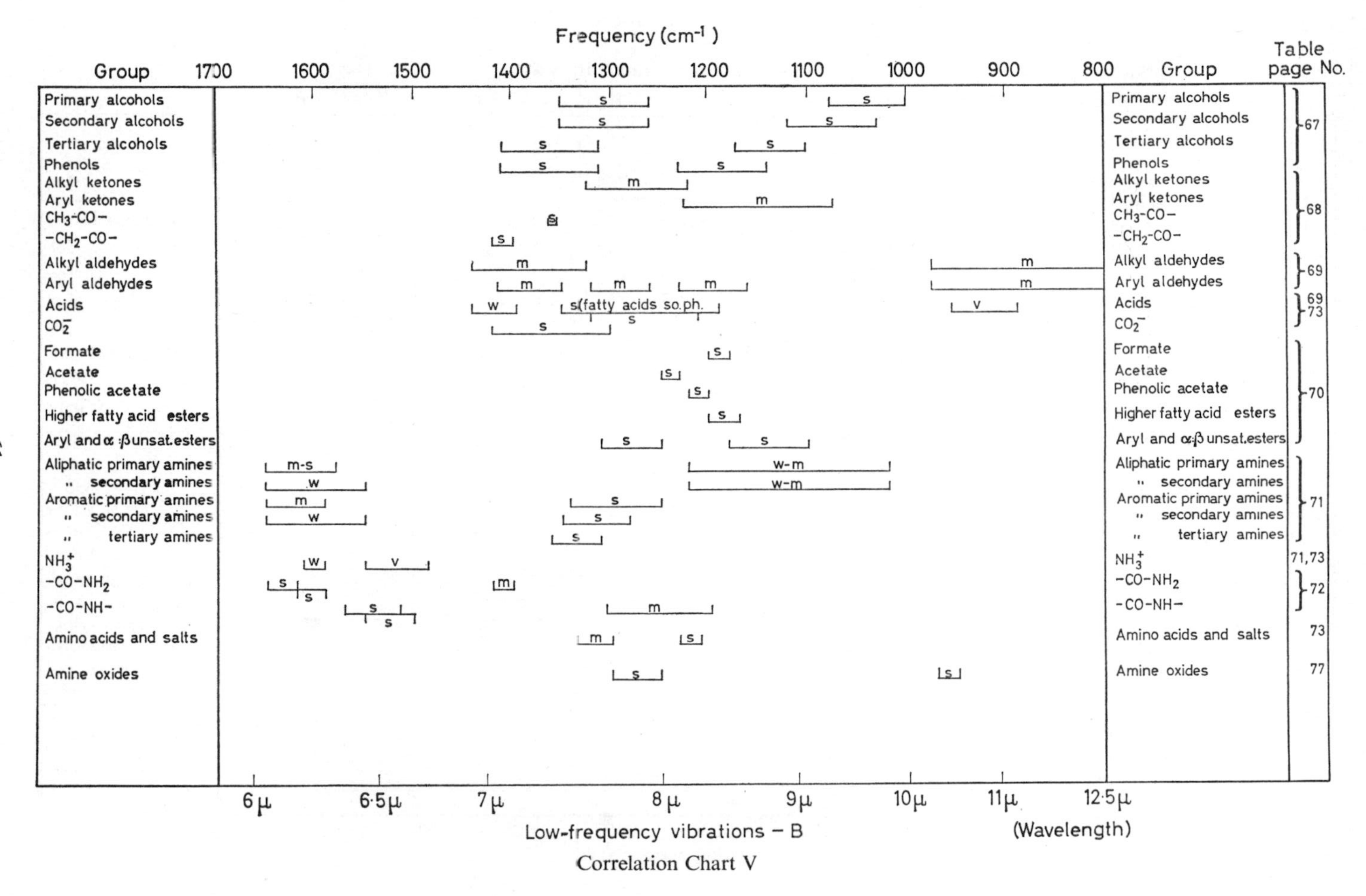

Low-frequency vibrations – B

Correlation Chart V

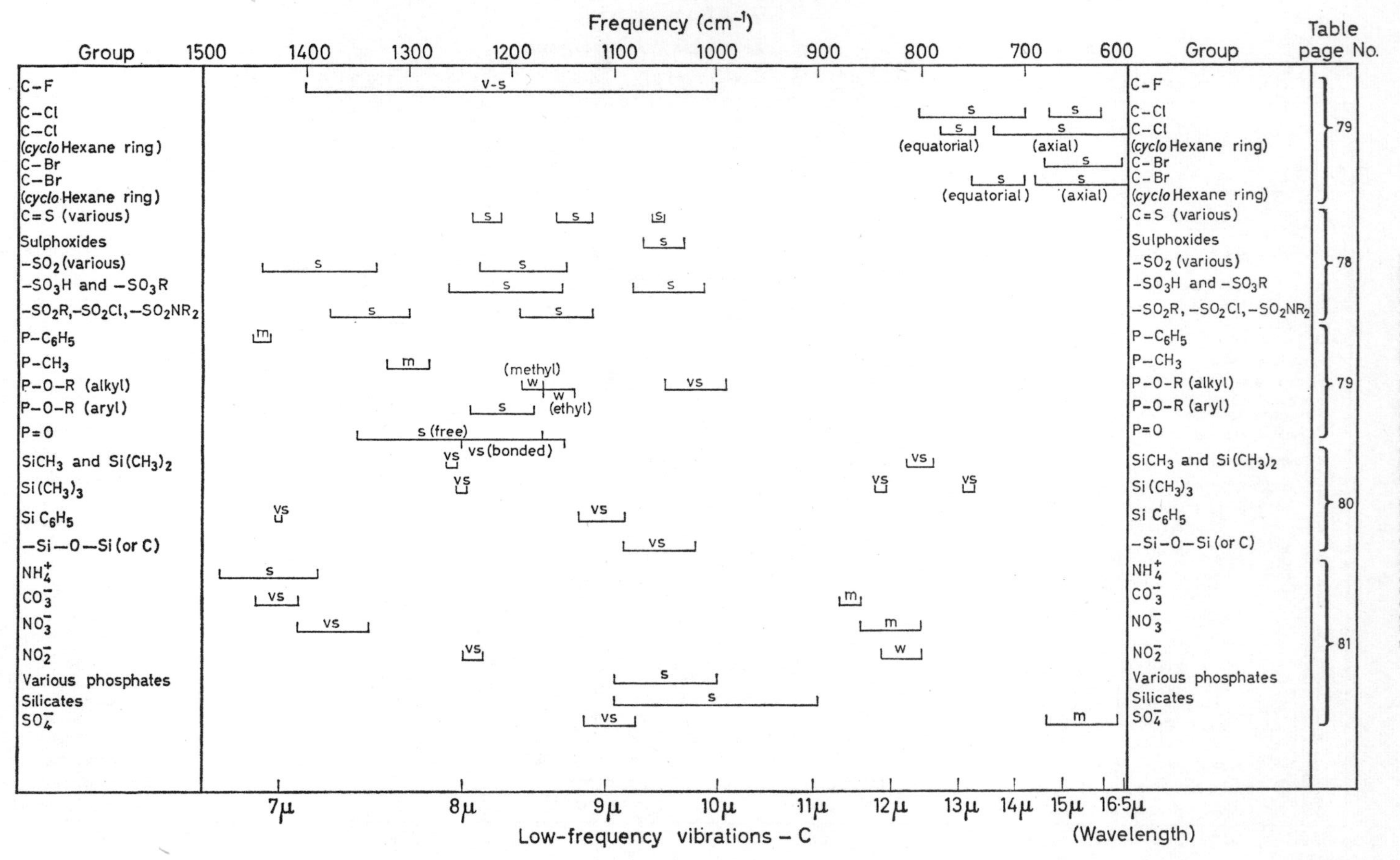

Correlation Chart VI

ALKANES

C—H *Stretching Vibrations*				
$—CH_3$	2,975–2,950	3·36–3·39	m.	The presence of several of these groups gives strong absorption
	2,885–2,860	3·47–3·50	m.	
$—CH_2—$	2,940–2,915	3·40–3·45	m.	
	2,870–2,845	3·49–3·52	m.	
$—CH_2—$ (*cyclo*propane)	3,080–3,040	3·25–3·29	v.	l.v.
—CH—	2,900–2,880	3·45–3·47	w.	l.v.
OCH_3, NCH_3 etc.				see ethers, amines etc.
C—H *Deformation Vibrations*				
$C—CH_3$	1,470–1,435	6·80–6·97	m.	asym. def.
	1,385–1,370	7·22–7·30	s.	sym. def.
$C(CH_3)_2$	1,385–1,380	7·22–7·25	s.	doublet of approx. equal int.
	1,370–1,365	7·30–7·33	s.	
$C(CH_3)_3$	1,395–1,385	7·17–7·22	m.	doublet int. ratio *ca.* 1:2
	1,365	7·33	s.	
$—CH_2—$	1,480–1,440	6·76–6·94	m.	CH_2 scissor
—CH—	*ca.* 1,340	*ca.* 7·46	w.	l.v.
Skeletal Vibrations				
$C(CH_3)_2$	1,175–1,165	8·51 –8·58	s.	
	1,170–1,140	8·55– 8·77	s.	
	840– 790	11·90–12·66	m.	l.v.
$C(CH_3)_3$	1,255–1,245	7·97– 8·03	s.	
	1,250–1,200	8·00– 8·33	s.	
$—(CH_2)_4$	750– 720	13·33–13·89	s.	
$—CH_2—$ (*cyclo*propane)	1,020–1,000	9·80–10·00	m.	l.v.

ALKENES

C=C *Stretching Vibrations*

non-conj. C=C	1,680–1,620	5·95–6·17	v.	
$CHR=CH_2$	1,645–1,640	6·08–6·10	v.	
$CHR_1=CHR_2$ (*cis*)	1,665–1,635	6·01–6·12	v.	
$CHR_1=CHR_2$ (*trans*)	1,675–1,665	5·97–6·00	v.	
$CR_1R_2=CH_2$	1,660–1,640	6·02–6·10	v.	
$CR_1R_2=CHR_3$	1,675–1,665	5·97–6·00	v.	
$CR_1R_2=CR_3R_4$	1,690–1,670	5·92–5·99	w.	l.v.
phenyl conj. C=C	*ca.* 1,625	*ca.* 6·16	s.	enh. int.
C=O or C=C conj. C=C	1,660–1,580	6·02–6·33	s.	cisoid form int. often more enh. than transoid

C—H *Stretching and Deformation Vibrations*

$CHR_1=CH_2$	3,040–3,010	3·29– 3·32	m.	CH str. (CHR_1)
	3,095–3,075	3·23– 3·25	m.	CH str. (CH_2)
	995– 985	10·05–10·15	m.	CH o.o.p. def.
	915– 905	10·93–11·05	s.	CH_2 o.o.p. def.
	1,850–1,800	5·41– 5·56	m.	overtone
	1,420–1,410	7·04– 7·09	w.	CH_2 i.p. def.
	1,300–1,290	7·69– 7·75	v.	CH i.p. def.
$CHR_1=CHR_2$ (*cis*)	3,040–3,010	3·29– 3·32	m.	CH str.
	1,420–1,400	7·04– 7·14	w.	CH i.p. def.
	730– 665	13·70–15·04	s.	CH o.o.p. def.
$CHR_1=CHR_2$ (*trans*)	3,040–3,010	3·29–3·32	m.	CH str.
	980– 960	10·20–10·42	s.	CH o.o.p. def.
	1,310–1,290	7·63– 7·75	w.	CH i.p. def.
$CR_1R_2=CH_2$	3,095–3,075	3·23– 3·25	m.	CH str.
	895– 885	11·17–11·30	s.	o.o.p. def.
	1,800–1,780	5·56– 5·62	m.	overtone
	1,420–1,410	7·04– 7·09	w.	CH_2 i.p. def.
$CR_1R_2=CHR_3$	3,040–3,010	3·29– 3·32	m.	CH str.
	850– 790	11·76–12·66	m.	CH o.o.p. def.

ALKYNES AND ALLENES

Alkynes

RC≡CH	3,310–3,300	3·02–3·03	m.	C—H str.
	2,140–2,100	4·67–4·76	w.	C≡C str.
$R_1C≡CR_2$	2,260–2,190	4·43–4·57	v.	C≡C str.

Allenes

C=C=C	1,970–1,950	5·08–5·13	m.	C≡C type str.
	ca. 1,060	*ca.* 9·43	m.	C—C type str.

AROMATIC HOMOCYCLIC COMPOUNDS

Stretching Vibrations

=C—H str.	3,080–3,030	3·25–3·30	w.-m.	multiple peaks may appear
C=C i.p. vib.	1,625–1,575	6·16–6·35	v.	usually close to 1,600 cm^{-1}
	1,525–1,475	6·56–6·78	v.	usually close to 1,500 cm^{-1}
	1,590–1,575	6·29–6·36	v.	strong band for conj. rings
	1,465–1,440	6·38–6·94	v.	

C—H *In-plane Deformations and Benzene Ring Substitution*

monosubstitution	1,175–1,125	8·51– 8·89	w.	
	1,110–1,070	9·01– 9·35	w.	
	1,070–1,000	9·35–10·00	w.	
1:2 disubstitution	1,225–1,175	8·17– 8·51	w.	
	1,125–1,090	8·89– 9·17	w.	
	1,070–1,000	9·35–10·00	w.	2 bands occur in this range
	1,000– 960	10·00–10·42	w.	
1:3 disubstitution	1,175–1,125	8·51– 8·89	w.	
	1,110–1,070	9·01– 9·35	w.	
	1,070–1,000	9·35–10·00	w.	
1:4 disubstitution	1,225–1,175	8·17– 8·51	w.	
	1,125–1,090	8·89– 9·17	w.	
	1,070–1,000	9·35–10·00	w.	2 bands occur in this range
1:2:3 trisubstitution	1,175–1,125	8·51– 8·89	w.	
	1,110–1,070	9·01– 9·35	w.	
	1,070–1,000	9·35–10·00	w.	
	1,000– 960	10·00–10·42	w.	
1:2:4 trisubstitution	1,225–1,175	8·17– 8·51	w.	
	1,175–1,125	8·51– 8·89	w.	
	1,125–1,090	8·89– 9·17	w.	
	1,070–1,000	9·35–10·00	w.	2 bands occur in this range
	1,000– 960	10·00–10·42	w.	
1:3:5 trisubstitution	1,175–1,125	8·51– 8·89	w.	
	1,070–1,000	9·35–10·00	w.	

C—H *Out-of-plane Deformations and Benzene Ring Substitution*

monosubstitution	770–730	12·99–13·70	s.	5 adj. free H atoms
	710–690	14·08–14·49	s.	5 ,, ,, ,,
1:2 disubstitution	770–735	12·99–13·61	s.	4 ,, ,, ,,
1:3 disubstitution	900–860	11·11–11·63	m.	1 free H atom
	810–750	12·35–13·33	s.	3 adj. free H atoms
	725–680	13·74–14·71	m.	l.v.; 3 ,, ,,
1:4 and 1:2:3:4 substitution	860–800	11·63–12·50	s.	2 ,, ,,
1:2:3 trisubstitution	800–770	12·50–12·99	s.	3 ,, ,,
	720–685	13·89–14·60	m.	l.v.; 3 ,, ,,
1:2:4 trisubstitution	860–800	11·63–12·50	s.	2 ,, ,,
	900–860	11·11–11·63	m.	1 free H atom
1:3:5 trisubstitution	900–860	11·11–11·63	m.	1 ,, ,,
	865–810	11·56–12·35	s.	1 ,, ,,
	730–675	13·70–14·81	s.	1 ,, ,,
1:2:3:5, 1:2:4:5 and 1:2:3:4:5 substitution	900–860	11·11–11·63	m.	1 ,, ,,

Benzene Ring Substitution Patterns of Summation Bands

Weak summation bands (overtones and combinations) of the CH out-of-plane deformation frequencies give absorption patterns in the range 2,000–1,650 cm^{-1} (5·00–6·06 μ), which are consistent and characteristic of the different substitutions of the benzene ring. Strong solutions are required to study these patterns [up to 20 times normal solution strengths (p. 23)]. Other bands occurring in this region, e.g. the strong C=C and C=O stretching fundamentals, mask the aromatic bands. Since the number of bands, their intensities and band shapes are more characteristic than absolute frequencies, no table is included here. These patterns are very useful in structural analysis and, though reference patterns are available[5], a preferred procedure is to prepare a set for each individual instrument.

ETHERS

C—O *Stretching Vibrations*

acyclic CH_2—O—CH_2	1,150–1,060	8·70– 9·43	s.	
C | C—C—O | C	920– 800	10·87–12·50	s.	
aryl and aralkyl =C—O—	1,270–1,230	7·87– 8·13	s.	
epoxides	1,260–1,240	7·94– 8·07	s.	i.b.
epoxides (*trans*)	950– 860	10·53–11·63	v.	l.v.
epoxides (*cis*)	865– 785	11·56–12·74	m.	l.v.
trimethylene oxides	980– 970	10·20–10·31	s.	
higher cyclic ethers	1,140–1,070	8·77– 9·35	s.	l.v.
—O—CH_2—O—	*ca.* 940	*ca.* 10·65	s.	l.v.
phthalans	915– 895	10·93–11·17	m.	

C—H *Stretching Vibrations*

—O—CH_3	2,830–2,815	3·53– 3·55	m.	
epoxide —CH—C— (O bridged)	3,050–2,990	3·28– 3·34	w.	
alkyl acetals, —CH(OCH_2—)$_2$	*ca.* 2,825	*ca.* 3·54	m.	
—O—CH_2—O—	*ca.* 2,780	*ca.* 3·60		
—CH=C—O— and —C(—)=CH—O—	3,150–3,050	3·18– 3·28	w.	

ALCOHOLS AND PHENOLS

O—H Stretching Vibrations

free OH	3,670–3,580	2·73–2·79	v.	sharp band
hydrogen bonded OH				
(a) intermolecular				
dimeric association	3,550–3,450	2·82–2·90	v.	sharp band } int. changes and frequency shifts on dilution
polymeric association	3,400–3,230	2·94–3·10	s.	broad band } int. changes and frequency shifts on dilution
(b) intramolecular	3,590–3,420	2·79–2·92	v.	sharp band } unaffected by dilution
(c) chelate compounds	3,200–1,700	3·13–5·88	w.	very broad band } unaffected by dilution
tropolones	*ca.* 3,100	*ca.* 3·23		
—OD	2,780–2,400	3·60–4·17	v.	O—D str.

C—O Stretching and O—H In-plane Deformations

primary alcohol	1,075–1,000	9·30–10·00	s.	l.v.
	1,350–1,260	7·40– 7·94	s.	l.v.
secondary alcohol	1,120–1,030	8·93– 9·71	s.	l.v.
	1,350–1,260	7·41– 7·94	s.	l.v.
tertiary alcohol	1,170–1,100	8·55– 9·09	s.	l.v.
	1,410–1,310	7·09– 7·63	s.	l.v.
phenols	1,230–1,140	8·13– 8·77	s.	l.v.
	1,410–1,310	7·09– 7·63	s.	l.v.

ACID HALIDES, ANHYDRIDES, AND SOME PEROXIDES

C—O Stretching Vibrations

anhydrides—cyclic	1,310–1,210	7·63–8·26	s.	
acyclic	1,175–1,045	8·51–9·57	s.	

C=O Stretching Vibrations

'5' ring anhydrides	1,870–1,830	5·35–5·46	s.	
	1,800–1,760	5·56–5·68	s.	
conj. '5' ring anhydrides	1,850–1,810	5·41–5·53	s.	
	1,795–1,740	5·57–5·75	s.	
acyclic anhydrides	1,840–1,800	5·44–5·56	s.	
	1,780–1,740	5·62–5·75	s.	
conj. acyclic anhydrides	1,820–1,780	5·50–5·62	s.	
	1,760–1,720	5·68–5·81	s.	
acyl peroxides	1,820–1,810	5·50–5·53	s.	
(R—CO—O—)$_2$	1,800–1,780	5·56–5·62	s.	
aroyl peroxides,	1,805–1,780	5·54–5·62	s.	
(R—CO—O—)$_2$	1,785–1,755	5·60–5·70	s.	
acid halides	1,815–1,785	5·51–5·60	s.	
conj. acid halides	1,800–1,770	5·56–5·65	s.	

—O—O— Vibration

peroxides	890–830	11·24–12·05	w.	l.v.

KETONES*†

C=O *Stretching Vibrations*

acyclic —CH_2—CO—CH_2—	1,725–1,700	5·80–5·88	s.	
α:β unsat. acyclic of '6' ring ketones	1,695–1,660	5·90–6·02	s.	
cross-conj. dienones	1,670–1,660	5·99–6·02	s.	
quinones—2 CO's in same ring	1,690–1,655	5·92–6·04	s.	
2 CO's in 2 rings	1,655–1,635	6·04–6·12	s.	
'4' ring ketones	1,780–1,760	5·62–5·68	s.	
'5' ring ketones	1,750–1,740	5·71–5·75	s.	
'6' ring ketones	1,720–1,700	5·81–5·88	s.	
'7' ring ketones	1,715–1,700	5·83–5·88	s.	
α-halogenated ketones†				
aryl ketones	1,700–1,680	5·88–5·95	s.	
diaryl ketones	1,670–1,660	5·99–6·02	s.	
—CO—CO—	1,730–1,710	5·78–5·85	s.	
—CO—CH_2—CO— (enolic)	1,640–1,535	6·10–6·52	s.	chelated, broad peak
—CO—C=C—OH (or NH_2)	1,640–1,535	6·10–6·52	s.	H bonding
ortho, —CO—C_6H_4—OH (or NH_2)	1,655–1,610	6·04–6·21	s.	H bonding
—CO—CH_2—CH_2—CO—	1,725–1,705	5·80–5·87	s.	
—CO—O—CH_2—CO—	1,745–1,725	5·73–5·80	s.	
tropolones	1,620–1,600	6·17–6·25	s.	

Other Vibrations

CH_3—CO—	1,360–1,355	7·35–7·38	s.	CH_3 def.
—CH_2—CO—	1,435–1,405	6·97–7·12	s.	CH_2 def.
alkyl ketones	1,325–1,215	7·55–8·23	m.	l.v.
aryl ketones	1,225–1,075	8·17–9·30	m.	l.v.
C=O	3,550–3,200	2·82–3·13	w.	C=O str. overtones

* For influence of physical state and medium on frequency of carbonyl bands see Part I, p. 32.
† For ketones, except those in which hydrogen bonding occurs, additive shifts of the original C=O stretching frequencies, and hence of the range limits given in the table, are observed for α substituents, as follows

α substituent	Frequency shift cm^{-1}	Wavelength shift μ	
α:β unsaturation	−30	+0·11	
*cyclo*propyl	−20	+0·07	l.v.
α halogen	+20	−0·07	in cyclic ketones only
αα′ halogens	+40	−0·15	equatorial halogen
αα halogens	+20	−0·07	causes +20 cm^{-1} shift

ALDEHYDES

$C{=}O$ *Stretching Vibrations*				
sat. aliphatic aldehydes	1,740–1,720	5·75–5·81	s.	
α:β-unsat. aldehydes	1,705–1,685	5·78–5·93	s.	
conj. polyene aldehydes	1,680–1,660	5·95–6·02	s.	
aryl aldehydes	1,715–1,695	5·83–5·90	s.	
—C(OH)=C—CHO	1,670–1,645	5·99–6·08	s.	intramolecular H bonding
C—H *Stretching and Deformation Vibrations*				
CHO	2,880–2,650	3·47– 3·77	w.-m.	C—H str. 2 bands may appear
	975– 780	10·26–12·82	w.	l.v.; C—H def.
Other Vibrations				
aliphatic aldehydes	1,440–1,325	6·94–7·55	m.	l.v.
aryl aldehydes	1,415–1,350	7·07–7·41	m.	l.v.
	1,320–1,260	7·58–7·94	m.	l.v.
	1,230–1,160	8·13–8·62	m.	l.v.

CARBOXYLIC ACIDS

O—H *Vibrations*				
free OH	3,550–3,500	2·82– 2·86	m.	O—H str.
bonded OH	3,300–2,500	3·00– 4·00	w.	broad band, O—H str.
all OH	955– 890	10·47–11·24	v.	o.o.p. def.
$C{=}O$ *Stretching Vibrations*				
sat. aliphatic acids	1,725–1,700	5·80–5·88	s.	all acids examined as dimers in so. ph. or liq. ph.
α:β-unsat. acids	1,715–1,680	5·83–5·95	s.	
aryl acids	1,700–1,680	5·88–5·95	s.	
intramolecular H bonded acids	1,680–1,650	5·95–6·06	s.	
α-halogeno acids	1,740–1,715	5·75–5·83	s.	
Other Vibrations				
solid fatty acids	1,350–1,180	7·40–8·48	w.	CH_2 vib., characteristic band patterns
CO_2H	1,440–1,395	6·94–7·17	w.	combination band of C—O str. and OH i.p. def.
	1,320–1,210	7·58–8·26	s.	
carboxylate ion CO_2^-	1,610–1,550	6·21–6·45	s.	asym. str.
	1,420–1,300	7·04–7·69	m.	sym. str.

ESTERS AND LACTONES*

C=O *Stretching Vibrations*

sat. aliphatic esters	1,750–1,735	5·71–5·76	s.	
α:β unsat. and aryl esters*	1,730–1,715	5·78–5·83	s.	
vinylic and phenolic esters	1,800–1,770	5·56–5·65	s.	
α-keto esters and α-diesters	1,755–1,740	5·70–5·75	s.	
enolic β-keto esters	1,655–1,635	6·04–6·12	s.	chelation
o-hydroxy (amino) benzoates, etc.	1,690–1,670	5·92–5·99	s.	chelation
γ-keto esters, non-enolic β-keto esters, and γ- (and higher) diesters	1,750–1,735	5·71–5·76	s.	
β-lactones	*ca.* 1,825	*ca.* 5·48	s.	
sat. γ-lactones	1,780–1,760	5·62–5·68	s.	
β:γ-unsat. γ-lactones	1,805–1,785	5·54–5·60	s.	
δ-lactones	1,750–1,735	5·71–5·76	s.	

C—O *Stretching Vibrations*

formates	1,200–1,180	8·33–8·48	s.	
acetates	1,250–1,230	8·00–8·13	s.	
vinylic and phenolic acetates	1,220–1,200	8·20–8·33	s.	
propionates and higher esters	1,200–1,170	8·33–8·55	s.	
esters of α:β unsat. aliphatic acids	1,310–1,250	7·63–8·00	s.	
	1,180–1,130	8·48–8·85	s.	
esters of aromatic acids	1,300–1,250	7·69–8·00	s.	
	1,150–1,100	8·70–9·09	s.	

* For α-substituted esters and lactones, other than those in which hydrogen bonding occurs, the following additive shifts of C=O stretching frequency (or wavelength) for individual compounds and range limits apply approximately:

α-substituent	Frequency shift cm^{-1}	Wavelength shift μ	
α:β double bond	−20	+0·07	
α-halogen	+20	−0·07	
αα-halogens	+20	−0·07	2 bands for soln. spec.

AMINES AND IMINES

N—H *Stretching Vibrations*				
primary amines	3,500–3,300	2·86–3·03	v.	2 bands appear in this range
secondary amines	3,500–3,300	2·86–3·03	v.	
imines	3,400–3,300	2·94–3·03	v.	l.v.
associated N—H	3,400–3,100	2·94–3·23	m.	
free N—D	2,600–2,400	3·85–4·15	v.	

N—H *Deformation Vibrations*				
primary amines	1,650–1,580	6·06–6·33	m.-s.	
secondary amines	1,650–1,550	6·06–6·45	w.	l.v.

C—N *Stretching Vibrations*				
aliphatic amines	1,220–1,020	8·20–9·80	w.-m.	l.v.
aromatic amines:				
primary	1,340–1,250	7·46–8·00	s.	
secondary	1,350–1,280	7·41–7·81	s.	
tertiary	1,360–1,310	7·35–7·63	s.	

Other Vibrations				
N-Methyl	2,820–2,760	3·55–3·62	m.-s.	C—H str.

CHARGED AMINE DERIVATIVES
(coordination complexes, amine hydrochlorides)

NH_3^+ *Stretching and Deformation Vibrations*				
NH_3^+	*ca.* 3,380	*ca.* 2·96	m.	NH_3^+ str. } values for soln. spectra only
	ca. 3,280	*ca.* 3·05	m.	NH_3^+ str. } values for soln. spectra only
	3,350–3,150	2·99–3·18	m.	NH_3^+ str., so. ph. spec., intermolecular H bonding, multiple bands may appear
	ca. 1,600	*ca.* 6·25	m.	asym. NH_3^+ def.
	ca. 1,300	*ca.* 7·69	m.	sym. NH_3^+ def.
	ca. 800	*ca.* 12·50	w.	NH_3^+ rocking

NH_2^+ *Vibrations*				
NH_2^+	*ca.* 2,700	*ca.* 3·70	s.	NH_2^+ str. vib.
	1,620–1,560	6·17–6·41	m.-s.	NII_2^+ dcf.
	ca. 800	*ca.* 12·50	w.	NH_2^+ rocking, l.v.

NH^+ *Vibrations*				
$C{=}NH^+$	2,500–2,325	4·00–4·30	s.	NH^+ str.
all NH^+	2,200–1,800	4·55–5·56	w.-m.	l.v., NH^+ str.

AMIDES

NH *Stretching Vibrations*

primary amides:				
free NH	3,540–3,480	2·83–2·88	s.	
	3,420–3,380	2·92–2·96	s.	
bonded NH	3,360–3,320	2·97–3·01	m.	
	3,220–3,180	3·11–3·15	m.	
secondary amides:				
free NH (*cis*)	3,440–3,420	2·91–2·93	s.	
free NH (*trans*)	3,460–3,440	2·89–2·91	s.	
bonded NH (*cis* and *trans*)	3,100–3,070	3·23–3·26	w.	
bonded NH (*cis*)	3,180–3,140	3·15–3·19	m.	
bonded NH (*trans*)	3,330–3,270	3·00–3·06	m.	

NH *Deformation Vibrations*

H bonded secondary amides	*ca.* 700	*ca.* 14·3		o.o.p. def., int. falls on dilution. Amide V band

C=O *Stretching Vibrations* (Amide—I band)

primary amides	*ca.* 1,690	*ca.* 5·92	s.	dil. soln. spec.
	ca. 1,650	*ca.* 6·06	s.	so. ph. spec.
secondary amides	1,700–1,665	5·88–6·01	s.	dil. soln. spec.
	1,680–1,630	5·95–6·14	s.	so. ph. spec.
tertiary amides	1,670–1,630	5·99–6·14	s.	dil. soln. or so. ph. spec.
simple β-lactams	1,760–1,730	5·68–5·78	s.	dil. soln. spec.
ring-fused β-lactams	1,780–1,770	5·62–5·65	s.	l.v., dil. soln. spec.
simple γ-lactams	*ca.* 1,700	*ca.* 5·88	s.	l.v.
ring-fused γ-lactams	1,750–1,700	5·71–5·88	s.	
larger-ring cyclic lactams	*ca.* 1,680	*ca.* 5·95	s.	dil. soln. spec.
ureas,				
—NH—CO—NH—	*ca.* 1,660	*ca.* 6·02	s.	
—CO—NH—CO—	1,790–1,720	5·59–5·81	s.	
	1,710–1,670	5·85–5·99	s.	
urethanes	1,735–1,700	5·76–5·88	s.	
carbamates	1,710–1,690	5·85–5·92	s.	l.v.

Combination Bands of NH *Deformation and* C—N *Stretching Vibrations*

primary amides	1,650–1,620	6·06–6·17	s.	so. ph. spec. } Amide II band
	1,620–1,590	6·17–6·31	s.	dil. soln. spec. } Amide II band
secondary acyclic amides	1,570–1,515	6·37–6·60	s.	so. ph. spec. } Amide II band
	1,550–1,510	6·45–6·62	s.	dil. soln. spec. } Amide II band
secondary amides	1,305–1,200	7·67–8·33	m.	l.v., i.p. combination, Amide III

Other Vibrations

primary amides	1,420–1,400	7·04– 7·14	m.	l.v.
secondary amides	770– 620	13·00–16·13	m.	l.v., Amide IV band
	630– 530	15·87–18·87	s.	l.v., Amide VI band

AMINO-ACIDS, AMIDO-ACIDS AND RELATED IONIC MOLECULES

Amino-acids

amino-acids containing an NH_2 group	3,130–3,030	3·20–3·30	m.	NH_3^+ str.
	1,660–1,610	6·02–6·21	w.	NH_3^+ def. Amino-acid I band
	1,550–1,485	6·45–6·73	v.	NH_3^+ def. Amino-acid II band
dicarboxylic α-amino-acids	1,755–1,720	5·70–5·81	s.	C=O str., unionized carboxyl
other dicarboxylic amino-acids	1,730–1,700	5·78–5·88	s.	C=O str., unionized carboxyl
dicarboxylic amino-acids	1,230–1,215	8·13–8·23	s.	C—O vib.
all amino acids	1,600–1,560	6·25–6·41	s.	ionized carboxyl, C=O str.
	2,760–2,530	3·62–3·95	w.	i.b., l.v.
	2,140–2,080	4·67–4·81	w.	NH_3^+ str., i.b., l.v.
	1,335–1,300	7·49–7·70	m.	i.b.

Amino-acid Salts H_2N—$(C)_n$—$CO_2^-M^+$

NH_2	3,400–3,200	2·94–3·13	m.	2 bands, NH_2 str.
CO_2^-	1,600–1,560	6·25–6·41	s.	ionized carboxyl C=O str.

Amino-acid Hydrochlorides H_3N^+—$(C)_n$—CO_2H Cl^-

NH_3^+	3,130–3,030	3·20–3·30	m.	NH_3^+ str., i.b.
	1,610–1,590	6·21–6·29	w.	NH_3^+ def.
	1,550–1,485	6·45–6·73	v.	NH_3^+ def.
α-amino-acid hydrochlorides	1,755–1,730	5·70–5.78	s.	C=O str.
other amino-acid hydrochlorides	1,730–1,700	5·78–5·88	s.	C=O str.
all amino-acid hydrochlorides	3,030–2,500	3·30– 4·0	w.	series of nearly continuous bands
	ca. 2,000	*ca.* 5·0	w.	
	1,335–1,300	7·49–7·70	m.	
	1,230–1,215	8·13–8·23	s.	C—O vib.

Amido-acids

NH	3,390–3,260	2·95–3·07	m.	N—H str.
α-amido-acids	1,725–1,695	5·80–5·90	s.	C=O str.
most amido-acids	2,640–2,360	3·79–4·24	w.	ib., l.v.
	1,945–1,835	5·14–5·45	w.	ib., l.v.
α-amido-acids	1,620–1,600	6·14–6·25	s.	Amide I band
other amido-acids	1,650–1,620	6·06–6·14	s.	Amide I band
all amido-acids	1,570–1,500	6·37–6·67	s.	Amido II band
	1,230–1,215	8·13–8·23	s.	C—O vib.

NON-AROMATIC UNSATURATED NITROGEN COMPOUNDS

C=N *Stretching Vibrations*				
acyclic C=N	1,690–1,635	5·92–6.12	v.	various oxazolones, oximes, oxazines, oxazolines, azomethines, etc.
acyclic α:β-unsat. C=N	1,665–1,630	6·01–6·14	v.	
cyclic α:β-unsat. C=N	1,660–1,480	6·02–6·76	v.	l.v., e.g. thiazoles

A=B=N *Allenic-type Stretching Vibrations*				
N=C=N	2,155–2,130	4·64–4·70	vs.	carbodiimides
N=C=O	2,275–2,240	4·40–4·46	vs.	isocyanates
[R—C=N=N]$^+$	2,280–2,260	4·39–4·43	s.	diazonium salts
—N=N=N	2,160–2,120	4·63–4·72	s.	azides, asym. str.
	1,350–1,180	7·41–8·48	w.	azides, l.v., sym. str

C≡N *Stretching Vibrations*				
sat. nitriles	2,260–2,240	4·43–4·46	w.-m.	
acyclic α:β-unsat. nitriles	2,235–2,215	4·47–4·52	s.	
aryl nitriles	2,240–2,220	4·46–4·51	m.-s.	
isonitriles	2,185–2,120	4·58–4·72	s.	l.v.
aryl nitriles	*ca.* 2,145	*ca.* 4·66	s.	l.v.

N=N *Stretching Vibrations*				
various azo compounds	1,630–1,575	6·14–6·35	v.	l.v.

NITROGEN HETEROCYCLES*

Pyridines and Quinolines

C—H *Stretching and Deformation Vibrations*

=C—H str.	3,070–3,020	3·26–3·31	s.	
ring CH deformations	*ca.* 1,200	*ca.* 8·33	s.	pyridines only
	1,100–1,000	9·09–10·00	s.	pyridines only
	900– 670	11·11–14·93	s.	*
	ca. 710	*ca.* 14·08	s.	pyridines only

C=C *and* C=N *Stretching Vibrations*

	1,650–1,580	6·06–6·33	m.	
	1,580–1,550	6·33–6·45	w.	i.b., l.v.
	1,510–1,480	6·62–6·76	m.	

Pyrimidines and Purines

C—H *Stretching and Deformation Vibrations*

=C—H str.	3,060–3,010	3·27– 3·32	s.	l.v.
ring CH deformations	1,000– 960	10·00–10·42	m.	
	825– 775	12·12–12·90	m.	

C=C *and* C=N *Stretching Vibrations*

	1,580–1,520	6·33–6·58	m.	l.v.

Pyrroles

N—H str.	3,440–3,400	2·91–2·94	m.	soln. spec.
C=C str.	*ca.* 1,565	*ca.* 6·39	v.	
	ca. 1,500	*ca.* 6·67	v.	

* Typical aromatic CH deformation bands occur for pyridines, quinolines and related heterocycles. As with benzenoid compounds the bands are characteristic of the number of adjacent free hydrogen atoms on the ring. In polyannular molecules each ring is considered separately. Thus γ-picoline absorbs at 800 cm^{-1}, similar to a *para*-substituted benzene, and *iso*quinoline shows bands for 1, 2 and 4 free adjacent ring hydrogen atoms.

Characteristic substitution patterns in the region 2,080–1,650 cm^{-1} (4·81–6·06 μ) have been observed for pyridines[6] (see p. 66).

COVALENT COMPOUNDS CONTAINING NITROGEN–OXYGEN BONDS

Oximes (R·C=NOH)

—NOH	3,650–3,500	2·74–2·86	v.	O—H str.

NO_2 *Vibrations, etc.—Nitro Compounds* (R·NO_2)

alkyl nitro compounds	920– 830	10·88–12·05	m.-s.	C—N vib., l.v.
primary and secondary nitro	1,565–1,545	6·39– 6·47	s.	asym. NO_2 str.
	1,385–1,360	7·22– 7·35	s.	sym. NO_2 str.
	1,380	7·25	m.	CH_2 def. in —CH_2—NO_2
tertiary nitro	1,545–1,530	6·47– 6·54	s.	asym. NO_2 str.
	1,360–1,340	7·35– 7·46	s.	sym. NO_2 str.
α:β-unsat. nitro	1,530–1,510	6·54– 6·62	s.	asym. NO_2 str.
	1,360–1,335	7·35– 7·49	s.	sym. NO_2 str.
α-halogeno nitro	1,580–1,570	6·33– 6·37	s.	asym. NO_2 str.
	1,355–1,340	7·38– 7·46	s.	sym. NO_2 str.
α:α dihalogenonitro	1,600–1,575	6·25– 6·35	s.	asym. NO_2 str.
	1,340–1,325	7·46– 7·55	s.	sym. NO_2 str.
aromatic nitro	1,550–1,510	6·45– 6·62	s.	asym. NO_2 str.
	1,365–1,335	7·33– 7·49	s.	sym. NO_2 str.
	860– 840	11·63–11·90	s.	C—N vib., l.v.
	ca. 750	*ca.* 13·33	s.	i.b., l.v.

NO_2 *Vibrations—Covalent Nitrates* (R·O·NO_2)

NO_2	1,655–1,610	6·04–6·21	s.	asym. NO_2 str.
	1,300–1,255	7·69–7·97	s.	sym. NO_2 str.

NO_2 *Vibrations—Nitramines* (R·N·NO_2)

sat. nitramines	1,585–1,530	6·31– 6·54	s.	asym. NO_2 str.
alkyl nitroguanidines	1,640–1,605	6·10– 6·23	s.	asym. NO_2 str.
aryl nitroguanidines and nitroureas	1,590–1,575	6·29– 6·35	s.	asym. NO_2 str.
all nitramines	1,300–1,260	7·69– 7·94	s.	sym. NO_2 str.
	790– 770	12·66–12·99	m.	l.v.

NO *Vibrations—Nitroso Compounds* (R·C·N=O)

aromatic nitroso	*ca.* 1,500	*ca.* 6·67	s.	N=O str., vap. ph. spec.
tert. aliphatic nitroso	*ca.* 1,550	*ca.* 6·45	s.	N=O str. } vap. ph. or soln. spec.
α-halogeno aliphatic nitroso	1,620–1,560	6·17–6 41	s.	N=O str. } vap. ph. or soln. spec.

COVALENT COMPOUNDS CONTAINING NITROGEN-OXYGEN BONDS—*continued*

NO *Vibrations—Nitrites* (R—O—N=O)

R—O—N=O *trans* form	1,680–1,650	5·95– 6·06	vs.	N=O str.
cis form	1,625–1,610	6·16– 6·21	vs.	N=O str.
R—O—N=O *trans* form	815– 750	12·27–13·33	s.	N—O str., l.v.
cis form	850– 810	11·76–12·35	s.	N—O str., l.v.
R—O—N=O *cis* form	690– 615	14·49–16·26	s.	O—N=O def., l.v.
trans form	625– 565	16·00–17·70	s.	O—N=O def., l.v.
R—O—N=O	3,360–3,220	2·98– 3·11	m.	N=O str. overtones

NO *Vibrations—Nitrosamines* (R·N·N=O)

N—N=O	1,500–1,480	6·67–6·76	s.	N=O str., vap. ph. spec.
	1,460–1,440	6·85–6·94	s.	N=O str., dil. soln. spec.
	ca. 1,050	*ca.* 9·52	s.	N—N str., l.v.
	ca. 660	*ca.* 15·15	s.	N—N=O def., l.v.

NO *Vibrations—Amine Oxides* (R·C·N→O)

pyridine and pyrimidine N-oxides	1,300–1,250	7·69– 8·00	m.-s.	N—O str., frequency varies widely with ring substituents
tert. aliphatic N-oxides	970– 950	10·31–10·53	s.	N—O str., l.v.
aryl nitrile N-oxides	*ca.* 1,370	*ca.* 7·30	s.	N—O str., l.v.

NO *Vibrations—Azoxy Compounds* (R·N·N→O)

	1,310–1,250	7·63–8·00	m.-s.	N—O str., l.v.

ORGANO-SULPHUR COMPOUNDS

C—S *Stretching Vibrations*

	705–570	14·18–17·54	w.	l.v.

C=S *Stretching Vibrations*

thioesters	*ca.* 1,675	*ca.* 5·97	s.	
—N—CS—N—, thioureas	1,430–1,130	6·99–8·85	s.	both bands associated with C—N vib. at *ca.* 1,300
—NH—CS—, thioamides	*ca.* 1,120	*ca.* 8·93	s.	
$(RS)_2$ C=S	1,060–1,050	9·43–9·52	s.	
$(RO)_2$ C=S	1,235–1,210	8·10–8·26	s.	
—C=C—C=S	1,155–1,140	8·66–8·77	s.	
$(aryl)_2$ C=S	1,230–1,215	8·13–8·23	s.	

S—H *Stretching Vibrations*

	2,590–2,550	3·86–3·92	w.	

S=O *Stretching Vibrations*

sat. or unsat. sulphoxides	1,070–1,030	9·35– 9·71	s.	so. ph. spec. 10–20 cm^{-1} lower
$R{\cdot}SO_2H$, sulphinic acids	*ca.* 1,090	*ca.* 9·17	s.	
$R_1{\cdot}SO_2R_2$, sulphinic esters	1,140–1,125	8·77– 8·89	s.	
$(RO)_2SO$, sulphites	1,220–1,170	8·20– 8·55	s.	
R_2SO_2, sulphones (sat. or unsat.)	1,160–1,120	8·62– 8·93	vs.	asym. str., so. ph. spec. 10–20 cm^{-1} lower
	1,350–1,300	7·41– 7·69	vs.	sym. str., so. ph. spec. 10–20 cm^{-1} lower
$R_1{\cdot}O{\cdot}SO_2{\cdot}R_2$, covalent sulphonate	1,420–1,330	7·04– 7·52	s.	
	1,200–1,145	8·33– 8·73	s.	
$(RO)_2SO_2$, covalent sulphate	1,440–1,350	6·94– 7·41	s.	
	1,230–1,150	8·13– 8·70	s.	
$R{\cdot}SO_3H$, sulphonic acid and $R{\cdot}SO_3^-$, ionic sulphonate	1,260–1,150	7·94– 8·70	s.	
	1,080–1,010	9·26– 9·90	s.	
	700– 600	14·29–16·67	s.	
$R{\cdot}SO_2Cl$, sulphonyl chlorides	1,375–1,340	7·27– 7·46	s.	
	1,190–1,160	8·40– 8·62	s.	
$R{\cdot}SO_2{\cdot}N$—sulphonamides	1,370–1,300	7·30– 7·69	s.	sym. str. } 10–20 cm^{-1} lower for so. ph. spec.
	1,180–1,140	8·48– 8·77	s.	asym. str. }

ORGANO-HALOGEN COMPOUNDS

C—X *Stretching Vibrations*

C—F monofluorinated compounds	1,110–1,000	9·01–10·00	s.	
C—F difluorinated compounds	1,250–1,050	8·00– 9·50	vs.	2 bands
C—F polyfluorinated compounds	1,400–1,100	7·14– 9·10	vs.	multiple bands
CF_3—CF_2	1,365–1,325	7·33– 7·55	s.	
C—Cl monochlorinated	750– 700	13·33–14·3	s.	
compounds	*ca.* 650	*ca.* 15·4	s.	for soln. spec. only
C—Cl equatorial	780– 750	12·80–13·33	s.	
C—Cl axial	730– 580	13·70–17·25	s.	
C—Cl polychlorinated compounds	800– 700	12·50–14·30	vs.-s.	l.v.
C—Br	*ca.* 650	*ca.* 15·4	s.	
	ca. 560	*ca.* 17·85	s.	for soln. spec. only
C—Br equatorial	750– 700	13·33–14·29	s.	
C—Br axial	690– 550	14·50–18·20	s.	
C—I	600– 500	16·67–20·00	s.	l.v.

C—X *Deformation Vibrations*

CF_3 aryl	1,330–1,310	7·52– 7·62	s.	sym. def.
	1,185–1,170	8·44– 8·55	s.	asym. def.
	1,150–1,130	8·70– 8·85	s.	asym. def.
CF_3—CF_2	745– 730	13·42–13·70	s.	

ORGANO-PHOSPHORUS COMPOUNDS

P—C *Vibrations, etc.*

P-phenyl	1,450–1,435	6·90–6·97	m.	
P-methyl	1,320–1,280	7·58–7·81	m.-s.	asym. CH_3 def.

P—H *Vibrations*

P—H str.	2,440–2,350	4·10– 4·26	m.	

P—O *Vibrations, etc.*

all P—O— alkyls	1,050– 990	9·52–10·10	vs.	
P—O— methyl	1,190–1,170	8·40– 8·55	w.	
P—O— ethyl	1,170–1,140	8·55– 8·77	w.	
P—O— aryl	1,240–1,180	8·07– 8·48	s.	
	ca. 1,030	*ca.* 9·71	w.	l.v.
P—OH	2,700–2,560	3·70– 3·90	w.	OH str., broad band, strong H bonding
P=O (free)	1,350–1,175	7·41– 8·51	s.	P=O str.
P=O (H bonded)	1,250–1,150	8·00– 8·70	vs.	P=O str.

ORGANO-SILICON COMPOUNDS

Si—C *Vibrations*				
Si—CH_3	1,260	7·94	vs.	sym. CH_3 def.
	ca. 800	*ca.* 12·50	vs.	Si—CH_3 str.
$Si(CH_3)_2$	1,260	7·94	vs.	sym. CH_3 def.
	815– 800	12·27–12·50	vs.	Si—CH_3 str.
$Si(CH_3)_3$	1,250	8·00	vs.	sym. CH_3 def.
	840	11·90	vs.	Si—CH_3 str.
	755	13·25	vs.	Si—CH_3 str.
Si-phenyl	1,430–1,425	6·99– 7·02	vs.	
	1,135–1,090	8·81– 9·17	vs.	
Si—H *Stretching Vibrations*				
Si—H str.	2,280–2,080	4·39– 4·81	vs.	
Si—O *Stretching Vibrations*				
Si—O—Si and Si—O—C	1,090–1,020	9·17– 9·80	vs.	Si—O str.

BORON COMPOUNDS

B—CH_3	1,460–1,405	6·85– 7·12		sym. def.
	1,320–1,280	7·58– 7·81		asym. def.
B—aryl	1,440–1,430	6·94– 6·99	s.	
B···· H—B	2,000–1,600	5·00– 6·25	v.	several bands may appear
BH_2	1,205–1,140	8·30– 8·77		i.p. def.
	975– 945	10·26–10·58		o.o.p. def.
	ca. 2,200	*ca.* 4·50		B—H str., double band
B—H	2,220–1,600	4·51– 6·25		several bands
	ca. 2,200	*ca.* 4·50		B—H str., single band
B—O	1,350–1,310	7·41– 7·63	s.	B—O str.
B—N	1,380–1,330	7·25– 7·52	s.	B—N str.
B—Cl	910– 890	10·99–11·24	s.	B—Cl. str.

INORGANIC IONS, Etc.

AsO_4^{3-}	*ca.* 800	*ca.* 12·50	s.	
AsF_6^-	705– 690	14·18–14·49	vs.	
BH_4^-	2,400–2,200	4·17– 4·55	s.	1 or more bands
	1,130–1,040	8·85– 9·62	s.	
BF_4^-	*ca.* 1,060	*ca.* 9·43	vs.	
	ca. 1,030	*ca.* 9·71	vs.	
BrO_3^-	810– 790	12·35–12·66	vs.	
CO_3^{2-}	1,450–1,410	6·90– 7·09	vs.	
	880– 800	11·36–12·50	m.	
HCO_3^-	1,420–1,400	7·04– 7·14	s.	
	1,000– 990	10·00–10·10	s.	
	840– 830	11·90–12·05	s.	
	705– 695	14·18–14·39	s.	
ClO_3^-	980– 930	10·20–10·75	vs.	
ClO_4^-	1,140–1,060	8·77– 9·43	vs.	broad absorption
CrO_4^{2-}	950– 800	10·53–12·50	s.	complex strong bands
$Cr_2O_7^{2-}$	950– 900	10·35–11·11	s.	
CN^-, CNO^-, and CNS^-	2,200–2,000	4·55– 5·00	s.	
CO	2,100–2,000	4·76– 5·00	s.	normal carbonyls
	ca. 1,830	*ca.* 5·46	s.	bridged carbonyls
HF_2^-	*ca.* 1,450	*ca.* 6·90	s.	
	ca. 1,230	*ca.* 8·13	s.	
IO_3^-	800– 700	12·50–14·29	s.	complex strong bands
MnO_4^-	920– 890	10·87–11·24	vs.	
	850– 840	11·76–11·90	m.	
NH_4^+	3,335–3,030	3·00– 3·30	vs.	
	1,485–1,390	6·73– 7·19	s.	
N_3^-	2,170–2,080	4·61– 4·81	s.	
	1,375–1,175	7·27– 8·51	w.	
NO_2^-	1,400–1,300	7·14– 7·69	s.	2 bands in complex nitrites
	1,250–1,230	8·00– 8·13	vs.	
	840– 800	11·90–12·50	w.	
NO_3^-	1,410–1,340	7·09– 7·46	vs.	
	860– 800	11·63–12·50	m.	
NO_2^+	1,410–1,370	7·09– 7·30	s.	
NO^+	2,370–2,230	4·22– 4·48	s.	
NO^+ (coordination comps.)	1,940–1,630	5·16– 6·14	s.	
NO^- (coordination comps.)	1,170–1,045	8·55– 9·57	s.	
NO (nitrosyl halides)	1,850–1,790	5·41– 5·59	s.	
PF_6^-	850– 840	11·76–11·90	vs.	
PO_4^{3-}, HPO_4^{2-}, and $H_2PO_4^-$	1,100– 950	9·09–10·53	s.	
$S_2O_3^{2-}$	1,660–1,620	6·02– 6.17	w.	
	1,000– 990	10·00–10·10	s.	
SO_4^{2-}	1,130–1,080	8·85– 9·26	vs.	
	680– 610	14·71–16·40	m.	
HSO_4^-	1,180–1,160	8·84– 8·62	s.	
	1,080–1,000	9·26–10·00	s.	
	880– 840	11·36–11·90	s.	
SO_3^{2-}	*ca.* 1,100	*ca.* 9·09	v.	l.v.
SeO_4^{2-}	*ca.* 830	*ca.* 12·05	s.	
SiF_6^{2-}	*ca.* 725	*ca.* 13·79	s.	
all silicates	1,100– 900	9·09–11·11	s.	
UO_2^{2+}	940– 900	10·64–11·11	s.	

RECIPROCALS

SUBTRACT

	0	1	2	3	4	5	6	7	8	9	1	2	3	4	5	6	7	8	9
1·0	1·0000	·9901	·9804	·9709	·9615	·9524	·9434	·9346	·9259	·9174	9	18	27	36	45	55	64	73	82
1·1	·9091	·9009	·8929	·8850	·8772	·8696	·8621	·8547	·8475	·8403	8	15	23	30	38	45	53	61	68
1·2	·8333	·8264	·8197	·8130	·8065	·8000	·7937	·7874	·7813	·7752	6	13	19	26	32	38	45	51	58
1·3	·7692	·7634	·7576	·7519	·7463	·7407	·7353	·7299	·7246	·7194	5	11	16	22	27	33	38	44	49
1·4	·7143	·7092	·7042	·6993	·6944	·6897	·6849	·6803	·6757	·6711	5	10	14	19	24	29	33	38	43
1·5	·6667	·6623	·6579	·6536	·6494	·6452	·6410	·6369	·6329	·6289	4	8	13	17	21	25	29	33	38
1·6	·6250	·6211	·6173	·6135	·6098	·6061	·6024	·5988	·5952	·5917	4	7	11	15	18	22	26	29	33
1·7	·5882	·5848	·5814	·5780	·5747	·5714	·5682	·5650	·5618	·5587	3	7	10	13	16	20	23	26	30
1·8	·5556	·5525	·5495	·5464	·5435	·5405	·5376	·5348	·5319	·5291	3	6	9	12	15	18	20	23	26
1·9	·5263	·5236	·5208	·5181	·5155	·5128	·5102	·5076	·5051	·5025	3	5	8	11	13	16	18	21	24
2·0	·5000	·4975	·4950	·4926	·4902	·4878	·4854	·4831	·4808	·4785	2	5	7	10	12	14	17	19	21
2·1	·4762	·4739	·4717	·4695	·4673	·4651	·4630	·4608	·4587	·4566	2	4	7	9	11	13	15	17	20
2·2	·4545	·4525	·4505	·4484	·4464	·4444	·4425	·4405	·4386	·4367	2	4	6	8	10	12	14	16	18
2·3	·4348	·4329	·4310	·4292	·4274	·4255	·4237	·4219	·4202	·4184	2	4	5	7	9	11	13	14	16
2·4	·4167	·4149	·4132	·4115	·4098	·4082	·4065	·4049	·4032	·4016	2	3	5	7	8	10	12	13	15
2·5	·4000	·3984	·3968	·3953	·3937	·3922	·3906	·3891	·3876	·3861	2	3	5	6	8	9	11	12	14
2·6	·3846	·3831	·3817	·3802	·3788	·3774	·3759	·3745	·3731	·3717	1	3	4	6	7	8	10	11	13
2·7	·3704	·3690	·3676	·3663	·3650	·3636	·3623	·3610	·3597	·3584	1	3	4	5	7	8	9	11	12
2·8	·3571	·3559	·3546	·3534	·3521	·3509	·3497	·3484	·3472	·3460	1	2	4	5	6	7	9	10	11
2·9	·3448	·3436	·3425	·3413	·3401	·3390	·3378	·3367	·3356	·3344	1	2	3	5	6	7	8	9	10
3·0	·3333	·3322	·3311	·3300	·3289	·3279	·3268	·3257	·3247	·3236	1	2	3	4	5	6	7	9	10
3·1	·3226	·3215	·3205	·3195	·3185	·3175	·3165	·3155	·3145	·3135	1	2	3	4	5	6	7	8	9
3·2	·3125	·3115	·3106	·3096	·3086	·3077	·3067	·3058	·3049	·3040	1	2	3	4	5	6	7	8	9
3·3	·3030	·3021	·3012	·3003	·2994	·2985	·2976	·2967	·2959	·2950	1	2	3	4	4	5	6	7	8
3·4	·2941	·2933	·2924	·2915	·2907	·2899	·2890	·2882	·2874	·2865	1	2	3	3	4	5	6	7	8
3·5	·2857	·2849	·2841	·2833	·2825	·2817	·2809	·2801	·2793	·2786	1	2	2	3	4	5	6	6	7
3·6	·2778	·2770	·2762	·2755	·2747	·2740	·2732	·2725	·2717	·2710	1	2	2	3	4	5	5	6	7
3·7	·2703	·2695	·2688	·2681	·2674	·2667	·2660	·2653	·2646	·2639	1	1	2	3	4	4	5	6	6
3·8	·2632	·2625	·2618	·2611	·2604	·2597	·2591	·2584	·2577	·2571	1	1	2	3	3	4	5	5	6
3·9	·2564	·2558	·2551	·2545	·2538	·2532	·2525	·2519	·2513	·2506	1	1	2	3	3	4	4	5	6
4·0	·2500	·2494	·2488	·2481	·2475	·2469	·2463	·2457	·2451	·2445	1	1	2	2	3	4	4	5	5
4·1	·2439	·2433	·2427	·2421	·2415	·2410	·2404	·2398	·2392	·2387	1	1	2	2	3	3	4	5	5
4·2	·2381	·2375	·2370	·2364	·2358	·2353	·2347	·2342	·2336	·2331	1	1	2	2	3	3	4	4	5
4·3	·2326	·2320	·2315	·2309	·2304	·2299	·2294	·2288	·2283	·2278	1	1	2	2	3	3	4	4	5
4·4	·2273	·2268	·2262	·2257	·2252	·2247	·2242	·2237	·2232	·2227	1	1	2	2	3	3	4	4	5
4·5	·2222	·2217	·2212	·2208	·2203	·2198	·2193	·2188	·2183	·2179	0	1	1	2	2	3	3	4	4
4·6	·2174	·2169	·2165	·2160	·2155	·2151	·2146	·2141	·2137	·2132	0	1	1	2	2	3	3	4	4
4·7	·2128	·2123	·2119	·2114	·2110	·2105	·2101	·2096	·2092	·2088	0	1	1	2	2	3	3	4	4
4·8	·2083	·2079	·2075	·2070	·2066	·2062	·2058	·2053	·2049	·2045	0	1	1	2	2	3	3	3	4
4·9	·2041	·2037	·2033	·2028	·2024	·2020	·2016	·2012	·2008	·2004	0	1	1	2	2	2	3	3	4
5·0	·2000	·1996	·1992	·1988	·1984	·1980	·1976	·1972	·1969	·1965	0	1	1	2	2	2	3	3	4
5·1	·1961	·1957	·1953	·1949	·1946	·1942	·1938	·1934	·1931	·1927	0	1	1	2	2	2	3	3	3
5·2	·1923	·1919	·1916	·1912	·1908	·1905	·1901	·1898	·1894	·1890	0	1	1	1	2	2	3	3	3
5·3	·1887	·1883	·1880	·1876	·1873	·1869	·1866	·1862	·1859	·1855	0	1	1	1	2	2	3	3	3
5·4	·1852	·1848	·1845	·1842	·1838	·1835	·1832	·1828	·1825	·1821	0	1	1	1	2	2	2	3	3
	0	1	2	3	4	5	6	7	8	9	1	2	3	4	5	6	7	8	9

RECIPROCALS

SUBTRACT

	0	1	2	3	4	5	6	7	8	9	1	2	3	4	5	6	7	8	9
5·5	·1818	·1815	·1812	·1808	·1805	·1802	·1799	·1795	·1792	·1789	0	1	1	1	2	2	2	3	3
5·6	·1786	·1783	·1779	·1776	·1773	·1770	·1767	·1764	·1761	·1757	0	1	1	1	2	2	2	3	3
5·7	·1754	·1751	·1748	·1745	·1742	·1739	·1736	·1733	·1730	·1727	0	1	1	1	2	2	2	2	3
5·8	·1724	·1721	·1718	·1715	·1712	·1709	·1706	·1704	·1701	·1698	0	1	1	1	1	2	2	2	3
5·9	·1695	·1692	·1689	·1686	·1684	·1681	·1678	·1675	·1672	·1669	0	1	1	1	1	2	2	2	3
6·0	·1667	·1664	·1661	·1658	·1656	·1653	·1650	·1647	·1645	·1642	0	1	1	1	1	2	2	2	3
6·1	·1639	·1637	·1634	·1631	·1629	·1626	·1623	·1621	·1618	·1616	0	1	1	1	1	2	2	2	2
6·2	·1613	·1610	·1608	·1605	·1603	·1600	·1597	·1595	·1592	·1590	0	1	1	1	1	2	2	2	2
6·3	·1587	·1585	·1582	·1580	·1577	·1575	·1572	·1570	·1567	·1565	0	0	1	1	1	1	2	2	2
6·4	·1563	·1560	·1558	·1555	·1553	·1550	·1548	·1546	·1543	·1541	0	0	1	1	1	1	2	2	2
6·5	·1538	·1536	·1534	·1531	·1529	·1527	·1524	·1522	·1520	·1517	0	0	1	1	1	1	2	2	2
6·6	·1515	·1513	·1511	·1508	·1506	·1504	·1502	·1499	·1497	·1495	0	0	1	1	1	1	2	2	2
6·7	·1493	·1490	·1488	·1486	·1484	·1481	·1479	·1477	·1475	·1473	0	0	1	1	1	1	2	2	2
6·8	·1471	·1468	·1466	·1464	·1462	·1460	·1458	·1456	·1453	·1451	0	0	1	1	1	1	2	2	2
6·9	·1449	·1447	·1445	·1443	·1441	·1439	·1437	·1435	·1433	·1431	0	0	1	1	1	1	1	2	2
7·0	·1429	·1427	·1425	·1422	·1420	·1418	·1416	·1414	·1412	·1410	0	0	1	1	1	1	1	2	2
7·1	·1408	·1406	·1404	·1403	·1401	·1399	·1397	·1395	·1393	·1391	0	0	1	1	1	1	1	2	2
7·2	·1389	·1387	·1385	·1383	·1381	·1379	·1377	·1376	·1374	·1372	0	0	1	1	1	1	1	2	2
7·3	·1370	·1368	·1366	·1364	·1362	·1361	·1359	·1357	·1355	·1353	0	0	1	1	1	1	1	2	2
7·4	·1351	·1350	·1348	·1346	·1344	·1342	·1340	·1339	·1337	·1335	0	0	1	1	1	1	1	1	2
7·5	·1333	·1332	·1330	·1328	·1326	·1325	·1323	·1321	·1319	·1318	0	0	1	1	1	1	1	1	2
7·6	·1316	·1314	·1312	·1311	·1309	·1307	·1305	·1304	·1302	·1300	0	0	1	1	1	1	1	1	2
7·7	·1299	·1297	·1295	·1294	·1292	·1290	·1289	·1287	·1285	·1284	0	0	0	1	1	1	1	1	1
7·8	·1282	·1280	·1279	·1277	·1276	·1274	·1272	·1271	·1269	·1267	0	0	0	1	1	1	1	1	1
7·9	·1266	·1264	·1263	·1261	·1259	·1258	·1256	·1255	·1253	·1252	0	0	0	1	1	1	1	1	1
8·0	·1250	·1248	·1247	·1245	·1244	·1242	·1241	·1239	·1238	·1236	0	0	0	1	1	1	1	1	1
8·1	·1235	·1233	·1232	·1230	·1229	·1227	·1225	·1224	·1222	·1221	0	0	0	1	1	1	1	1	1
8·2	·1220	·1218	·1217	·1215	·1214	·1212	·1211	·1209	·1208	·1206	0	0	0	1	1	1	1	1	1
8·3	·1205	·1203	·1202	·1200	·1199	·1198	·1196	·1195	·1193	·1192	0	0	0	1	1	1	1	1	1
8·4	·1190	·1189	·1188	·1186	·1185	·1183	·1182	·1181	·1179	·1178	0	0	0	1	1	1	1	1	1
8·5	·1176	·1175	·1174	·1172	·1171	·1170	·1168	·1167	·1166	·1164	0	0	0	1	1	1	1	1	1
8·6	·1163	·1161	·1160	·1159	·1157	·1156	·1155	·1153	·1152	·1151	0	0	0	1	1	1	1	1	1
8·7	·1149	·1148	·1147	·1145	·1144	·1143	·1142	·1140	·1139	·1138	0	0	0	1	1	1	1	1	1
8·8	·1136	·1135	·1134	·1133	·1131	·1130	·1129	·1127	·1126	·1125	0	0	0	1	1	1	1	1	1
8·9	·1124	·1122	·1121	·1120	·1119	·1117	·1116	·1115	·1114	·1112	0	0	0	1	1	1	1	1	1
9·0	·1111	·1110	·1109	·1107	·1106	·1105	·1104	·1103	·1101	·1100	0	0	0	1	1	1	1	1	1
9·1	·1099	·1098	·1096	·1095	·1094	·1093	·1092	·1091	·1089	·1088	0	0	0	0	1	1	1	1	1
9·2	·1087	·1086	·1085	·1083	·1082	·1081	·1080	·1079	·1078	·1076	0	0	0	0	1	1	1	1	1
9·3	·1075	·1074	·1073	·1072	·1071	·1070	·1068	·1067	·1066	·1065	0	0	0	0	1	1	1	1	1
9·4	·1064	·1063	·1062	·1060	·1059	·1058	·1057	·1056	·1055	·1054	0	0	0	0	1	1	1	1	1
9·5	·1053	·1052	·1050	·1049	·1048	·1047	·1046	·1045	·1044	·1043	0	0	0	0	1	1	1	1	1
9·6	·1042	·1041	·1040	·1038	·1037	·1036	·1035	·1034	·1033	·1032	0	0	0	0	1	1	1	1	1
9·7	·1031	·1030	·1029	·1028	·1027	·1026	·1025	·1024	·1022	·1021	0	0	0	0	1	1	1	1	1
9·8	·1020	·1019	·1018	·1017	·1016	·1015	·1014	·1013	·1012	·1011	0	0	0	0	1	1	1	1	1
9·9	·1010	·1009	·1008	·1007	·1006	·1005	·1004	·1003	·1002	·1001	0	0	0	0	0	1	1	1	1
	0	1	2	3	4	5	6	7	8	9	1	2	3	4	5	6	7	8	9

RECIP.

REFERENCES—PART II

[1] BELLAMY, L. J., *Infra-red Spectra of Complex Molecules* (2nd edn); Methuen: London, 1958.
[2] JONES, R. N. and SANDORFY, C., *Chemical Applications of Spectroscopy*; Interscience: New York, 1956.
[3] COLTHUP, N. B., 'Spectra–structure Correlations in the Infra-red Region.' *J. opt. Soc. Amer.*, 1950, **40**, 397.
[4] NAKANISHI, K., *Infra-red Absorption Spectroscopy—Practical*; Holden-Day Inc.: San Francisco, 1962.
[5] YOUNG, C. W., DUVALL, R. B. and WRIGHT, N., *Anal. Chem.*, 1951, **23**, 709.
[6] COOK, G. L. and CHURCH, F. M., *J. phys. Chem.*, 1957, **61**, 458.

INDEX

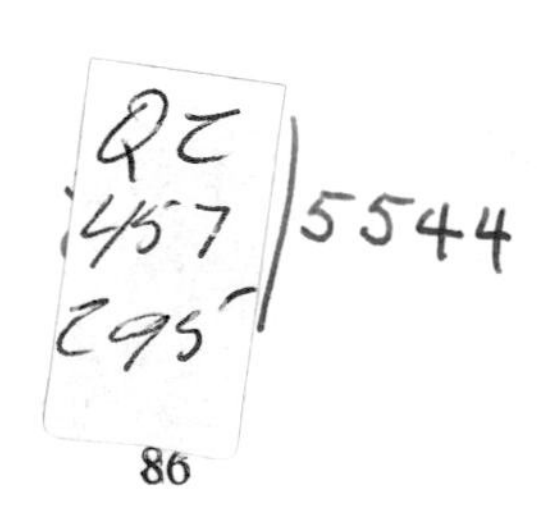